GREEN
ARCHITECTURE

McGRAW-HILL'S GREENSOURCE SERIES

Attmann
Green Architecture: Advanced Technologies and Materials

Gevorkian
Solar Power in Building Design: The Engineer's Complete Design Resource
Alternative Energy Systems in Building Design

GreenSource: The Magazine of Sustainable Design
Emerald Architecture: Case Studies in Green Building

Haselbach
The Engineering Guide to LEED—New Construction: Sustainable Construction for Engineers

Luckett
Green Roof Construction and Maintenance

Melaver and Mueller (eds.)
The Green Building Bottom Line: The Real Cost of Sustainable Building

Nichols and Laros
Inside the Civano Project: A Case Study of Large-Scale Sustainable Neighborhood Development

Yudelson
Green Building Through Integrated Design
Greening Existing Buildings

About *GreenSource*

A mainstay in the green building market since 2006, *GreenSource* magazine and GreenSourceMag.com are produced by the editors of McGraw-Hill Construction, in partnership with editors at BuildingGreen, Inc., with support from the United States Green Building Council. *GreenSource* has received numerous awards, including American Business Media's 2008 Neal Award for Best Website and 2007 Neal Award for Best Start-up Publication, and FOLIO magazine's 2007 Ozzie Awards for "Best Design, New Magazine" and "Best Overall Design." Recognized for responding to the needs and demands of the profession, *GreenSource* is a leader in covering noteworthy trends in sustainable design and best practice case studies. Its award-winning content will continue to benefit key specifiers and buyers in the green design and construction industry through the books in the *GreenSource* Series.

About McGraw-Hill Construction

McGraw-Hill Construction, part of The McGraw-Hill Companies (NYSE: MHP), connects people, projects, and products across the design and construction industry. Backed by the power of Dodge, Sweets, *Engineering News-Record* (*ENR*), *Architectural Record*, *GreenSource*, *Constructor*, and regional publications, the company provides information, intelligence, tools, applications, and resources to help customers grow their businesses. McGraw-Hill Construction serves more than 1,000,000 customers within the $4.6 trillion global construction community. For more information, visit www.construction.com.

About the International Code Council

The International Code Council (ICC) is a nonprofit membership association dedicated to protecting the health, safety, and welfare of people by creating better buildings and safer communities. The mission of ICC is to provide the highest quality codes, standards, products and services for all concerned with the safety and performance of the built environment. ICC is the publisher of the family of the International Codes® (I-Codes®), a single set of comprehensive and coordinated national model codes. This unified approach to building codes enhances safety, efficiency and affordability in the construction of buildings. The Code Council is also dedicated to innovation, sustainability and energy efficiency. Code Council subsidiary, ICC Evaluation Service, issues Evaluation Reports for innovative products and Reports of Sustainable Attributes Verification and Evaluation (SAVE).

Headquarters: 500 New Jersey Avenue, NW, 6th Floor, Washington, DC 20001-2070; *District Offices*: Birmingham, AL; Chicago. IL; Los Angeles, CA, 1-888-422-7233, www.iccsafe.org.

GREEN ARCHITECTURE
ADVANCED TECHNOLOGIES AND MATERIALS

OSMAN ATTMANN

New York   Chicago   San Francisco   Lisbon   London   Madrid
Mexico City   Milan   New Delhi   San Juan   Seoul
Singapore   Sydney   Toronto

The McGraw-Hill Companies

Cataloging-in-Publication Data is on file with the Library of Congress.

McGraw-Hill books are available at special quantity discounts to use as premiums and sales promotions, or for use in corporate training programs. To contact a representative please e-mail us at bulksales@mcgraw-hill.com.

Green Architecture: Advanced Technologies and Materials

Copyright © 2010 by Osman Attmann. All rights reserved. Printed in the United States of America. Except as permitted under the United States Copyright Act of 1976, no part of this publication may be reproduced or distributed in any form or by any means, or stored in a data base or retrieval system, without the prior written permission of the publisher.

1 2 3 4 5 6 7 8 9 0 DOC/DOC 0 1 4 3 2 1 0 9

ISBN 978- 0-07-162501-2
MHID 0-07-162501-1

Sponsoring Editor
Joy Bramble

Acquisitions Coordinator
Michael Mulcahy

Editorial Supervisor
David E. Fogarty

Project Manager
Preeti Longia Sinha, Glyph International Limited

Copy Editor
Bhavna Gupta, Glyph International Limited

Proofreader
Upendra Prasad, Glyph International Limited

Production Supervisor
Richard C. Ruzycka

Composition
Glyph International Limited

Art Director, Cover
Jeff Weeks

The pages within this book were printed on acid-free paper containing 100% postconsumer fiber.

Information contained in this work has been obtained by The McGraw-Hill Companies, Inc. ("McGraw-Hill") from sources believed to be reliable. However, neither McGraw-Hill nor its authors guarantee the accuracy or completeness of any information published herein, and neither McGraw-Hill nor its authors shall be responsible for any errors, omissions, or damages arising out of use of this information. This work is published with the understanding that McGraw-Hill and its authors are supplying information but are not attempting to render engineering or other professional services. If such services are required, the assistance of an appropriate professional should be sought.

About the Author

Osman Attmann (Denver, Colorado) is an architect and Associate Professor at the College of Architecture and Planning, University of Colorado. Professor Attmann, who has an M. Arch from the State University of New York and a Ph.D. from the Georgia Institute of Technology, publishes regularly on green architecture, emerging technologies, and sustainability. As a designer and co-developer of architectural smart-wall systems, Professor Attmann is professionally involved in a variety of advanced technology and green material research projects, has received several grants, and holds multiple patents. He is the Principal of Green-Tecture Studio: Architectural Design and Emerging Technologies research and advisory firm.

CONTENTS

PREFACE

Global warming is emerging as an issue of international significance. The rapid growth and increasing concentration of greenhouse gases worldwide is expected to contribute to climate change at a faster rate than previously recorded in the earth's history. Governments, industries, communities, and academic fields across the globe are racing to better understand the impact(s) of these climatic changes and their implications for ecological balance. Today, it is the responsibility of every academic field to help create sustainable environments by developing solutions for reducing energy and water consumption, and our dependence on nonrenewable resources.

This is especially imperative in the architecture and construction fields, where buildings account for nearly half of all greenhouse gas emissions, energy consumption, and raw material use around the globe. According to 2006 Energy Information Administration data, commercial and residential buildings account for 48 percent of the energy consumed, 76 percent of electricity used, and 15 percent of the total water consumed. Buildings also use 50 percent of the world's raw materials—many of which are nonrenewable resources—and they are responsible for 36 percent of all waste generated worldwide. These numbers are alarming and clearly identify two distinct problems in the building sector that must be taken seriously and addressed immediately.

The first problem is the nature and use of our existing technology, which is largely inefficient, outdated, and even in some cases obsolete. In the United States alone, the building sector uses nonrenewable and environmentally hazardous energy, emits 39 percent of all energy-related carbon dioxide, lacks the ability to retain its generated energy long enough to sustain indoor air quality levels, and fails in recycling and managing water and waste.

The second problem is the materials, mostly because of their compositional attributes, manufacturing, and fabrication methods. According to recent reports published by the U.S. Environmental Protection Agency (EPA, 2001), Department of Energy (Annual Energy Review, 2003, 2006) and Department of Commerce (General Housing Characteristics, 2000; Economic Census, 2002), existing building materials show significant deficiencies in most of the performance issues, such as durability, energy efficiency, amount of waste generated, toxicity, and potential for reuse.

Given the fact that inefficient buildings are partially responsible for these environmental changes, the following questions need to be asked: What is the role of architects and what can we do to address these problems? These questions are the main focus of the book, with an emphasis on utilizing advanced technologies and materials as a means to help remedy these serious issues. According to the U.S. Energy Information Administration forecast report (U.S. Annual Energy Outlook, 2006), by the year 2035, 75 percent of the existing building stock will be either replaced or renovated. This transformation over the next two decades represents a historic opportunity for architects, engineers, and the building community to develop and use new advanced materials and technologies in order to make these future buildings efficient, environmentally friendly, and sustainable.

This book intends to contribute to this historic transformation by introducing new technologies and materials that will promote and support this change.

The book is organized into eight chapters around the following themes:
Chapter 1 introduces the subject, providing an overview of the issues within our ecosystem. It concludes with a discussion of our responsibilities as architects, and addresses the relationship between architecture and technology.

Chapter 2 compares the concepts sustainability, ecology, and the green movement. Sustainability is of one the most widely used but poorly defined terms in architecture today. While the terms "sustainable," "ecological," and "green" are often used interchangeably to describe environmentally responsive architecture, in depth each term has its own sociopolitical connotations and agenda. This chapter defines these terms separately, investigates their similarities and differences, and explores their development over the years. It concludes by focusing on green architecture as a concept that best describes technologically advanced, economical, environmentally friendly, sustainable architecture.

Chapter 3 provides a brief history of the preservation, ecological, environmental, and green movements, as well as green architecture, including a summary of current Green Building Rating Systems.

Chapters 4 and 5 present existing, emerging, and future green technologies, as they apply to residential, commercial, and institutional buildings. It provides a complete introduction to the advanced technologies used to construct high-performance buildings. This chapter focuses on three major technological issues: (1) energy generation (i.e., biological, solar, geothermal, and hybrid systems); (2) energy retention (i.e., insulation, environmental controls, skins, building envelopes, passive methods, and zero-energy technologies); and (3) water and waste management (i.e., reuse, recycle, energy recovery, and disposal). Furthermore, this chapter emphasizes and encourages the reader to realize the ecological, economical, and design benefits of using green technologies.

Chapters 6 and 7 focus on green materials that make "zero-energy" green buildings. Building materials and construction activities consume three billion tons of raw materials each year. Using green building materials and products promotes conservation of dwindling nonrenewable resources internationally, and can also help reduce environmental impacts (e.g., extraction, transportation, processing, installation, and disposal) associated with traditional building industry materials. Accordingly, these chapters focus on several categories of green materials (such as biomaterials, composites, smart and nanomaterials) with the following benefits: (1) energy exchange and conservation; (2) environmental adaptation and reversibility; (3) environmental safety, recyclability, and renewability; (4) reduced maintenance/replacement costs over the life span of a building; and (5) greater design attributes and flexibility.

Chapter 8 takes an in-depth look at a broad range of relevant case studies in green architecture that are designed, operated, renovated, or reused in an ecological, resource-efficient manner. Selected projects represent both new construction and future projects for various building types, based on the earlier defined standards of green architecture (i.e., employing alternative energies, incorporating advanced technologies and materials, using efficient water- and waste-management techniques, and reducing the overall impact to the environment). Case studies are presented and analyzed in four major categories: (1) advanced green buildings; (2) active and passive solar buildings; (3) self-sufficient, off-the-grid modular and mobile systems; and (4) solar-decathlon competition projects since 2002. The last three categories have never been covered or discussed in other publications.

It is my hope that this book will help the reader to understand that green architecture is not only good for the environment, but offers new opportunities for creativity and innovation.

ACKNOWLEDGMENTS

This book was written with the support and assistance of many people. I am thankful to each one of them and I regret that I can only list a few in a one-page acknowledgement section.

First, I would like to thank my wife, Julianne, for her continuous support, constructive criticism, editing assistance, and for our endless brainstorming sessions that helped me to formulate my ideas. In addition, I would like to thank Marian Trautmann for her efficient and effective editing help, especially during tight deadlines.

I would like to thank my dean, Mark Gelernter, for his continuous support on this project. I am also grateful to my colleagues, who challenged my ideas and helped me to organize my thoughts throughout this project.

This book would not have been completed without the help of my CU seminar students, in both Denver and Boulder, some of whom deserve to be acknowledged personally: Jason Barnes, Alicia Bock, Timothy Barstad, Evan Brooks, Lori Dunn, Kendall Goodman, Brittany Hanna, Tanya Jimenez, Scott Knoll, Erik Kramer, Kathleen Mannis, Luke Martin, Amanda Martin, Luis Navarrete, Rachel Saunders, McClees Stephens, Steven Swanson, Tom Commerford, Bill Daher, and Lauren Watkins.

My editor, Joy Bramble, deserves a special acknow-ledgment for her continued belief in this book's completion.

Finally, my ultimate appreciation goes to my daughter Su, who constantly wondered why I had to work in my office instead of playing with her and watching the *Simpsons* at home, but patiently understood, nevertheless.

GREEN ARCHITECTURE: OVERVIEW

Introduction: Ecosystems and Natural Environments

An ecosystem is defined as a natural unit, consisting of all plants, animals, and microorganisms in an area functioning together, along with the nonliving factors of the area (Christopherson 1997). The term was first coined by A. G. Tansley in 1935 to encompass the interactions between biotic and abiotic components of the environment at a given site (Tansley 1935; Gorham 2006). The word biotic refers to the living components of an environment, where the actions of a variety of species affect the lives of fellow organisms. Abiotic factors are essentially nonliving components, such as temperature, light, moisture, air currents, etc. that equally affect the ecosystem (Lockwood and McKinney 2001; Buchs 2003; Saxena 2003; Gaston and Spicer 2004).

The term ecosystem was later redefined by Eugene Odum as any unit where this interaction between biotic and abiotic factors in a given area produces a flow of energy leading to a clearly defined trophic structure, biotic diversity, and material cycles (exchange of materials between living and nonliving parts) within the system (Odum 1971; Gorham 2006).

An ecosystem is not a single unified entity, constant in size. Our entire planet is covered with a variety of different, sometimes overlapping, and often interdependent ecosystems. A single lake, a neighborhood, or an entire region can be considered an ecosystem, while the term biome is used to refer to a major global ecosystem (Millennium Ecosystem Assessment 2005, *Ecosystems and Human Well-being: Current State and Trends*). Therefore an ecosystem can be as large as the entire globe, or as small as a city or a building (see Fig. 1.1).

While smaller ecosystems are part of the global ecosystem, the flow of energy for an ecosystem occurs on a more local level. Most ecosystems are *autotrophic,* which means that they capture sufficient energy to support their own requirements. For example, green plants convert solar energy to glucose, which is used for plant growth and other functions. In turn they provide energy to the rest of the living system. However, some ecosystems are *heterotrophic*, that is, unable to produce sufficient energy to meet the system's needs. The additional energy requirements must come from adjacent ecosystems. Thus, they can be regarded as autotrophic

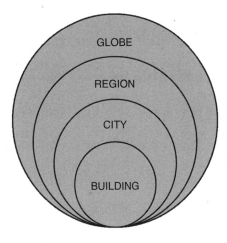

Figure 1.1   Ecosystems are a hierar-chy of systems, consisting of subsystems which make up parts of supersystems.

on a bioregional level, such as a river and its watershed (Wardle 2002; Newman and Jennings 2008).

The key to an ecosystem is interconnection and relationship. All the parts of an ecosystem are interrelated through a complex set of self-regulating cycles, feedback loops, and linkages between different parts of the food chain. If one part of an ecosystem is removed or disrupted, there are ripple effects throughout the system. The extent of the disturbance varies depending on the nature, scale, and duration of the disruption; on the relative significance of the part or parts affected; and on the resilience of the ecosystem (Chapin, Mooney and Chapin 2004; ANRC 2005; Ponting 2007; Krapivin and Varotsos 2008; Ostfeld, Keesing and Eviner 2008).

This complex, unique, and fragile relationship is quite vulnerable. It can be altered, even damaged, by various factors, whether cyclical, natural, or man-made. Examples of cyclical factors are solar flares and radiations (Hoyt and Schatten 1997; Carslaw, Harrison and Kirkby 2002; Salby and Callaghan 2004; Benestad 2006), the orbital inclination of the earth with the astronomical theory of accompanying climate change (Berger 2002; Svensmark 2007), and climate variations from geology, geochemistry, and paleontology (Saltzman 2001; Lovejoy and Hannah 2005). Natural factors refer to such events as earthquakes, vol-canic activities, floods, and fires (Kondratev and Galindo 1997; Cowie 2007). Most of these factors are out of our control and run their natural course, regardless of what we do.

Human Impact on the Natural Environment

Man-made factors are controllable, and our actions have a great impact on the ecosystem. Throughout history, man's interaction with nature has created disruptive and damaging effects, whether through the generation of energy, the creation of artificial landscapes, the construction of buildings, excavations, or soil cultivation. Today, our pressure on natural environments and the magnitude of the disruption of ecosystems is greater than ever. According to the 2005 Millennium Ecosystem Assessment (MEA) study, the health of the world's ecosystems is in significant decline (Millennium Ecosystem Assessment 2005, *Ecosystems and Human Well-being: Synthesis*). Whether one considers the supply of fresh water and food, or the regulation of climate and air quality, the study found that the global ecosystems showed a 62 percent decline over the course of the last four decades (see Fig. 1.2).

Just a few of the problems resulting from this decline are: 40 percent of the world's coral reefs have been lost or degraded, water withdrawals from rivers and lakes have doubled since 1960, the atmospheric concentration of carbon dioxide has jumped 19 percent since 1959, and extinction of species has increased as much as 100 times over the typical rate seen across earth's history (Millennium Ecosystem Assessment 2005, *Ecosystems and Human Well-being: Synthesis*) (see Fig. 1.3.).

So what does this mean to us? The decline of these systems brings about an increasing risk of disruptive and potentially irreversible changes, such as regional climate shifts, the emergence of new diseases, and the formation of dead zones in coastal waters (Millennium Ecosystem Assessment 2005). Another unforeseen consequence of this decline is the exacerbation of poverty among the two-thirds of the world's population who desperately rely on the resources formerly produced by collapsing local ecosystems.

Furthermore, two separate reports, published by MEA 2005 (Millennium Ecosystem Assessment 2005,) and the World Wide Fund for Nature (Living beyond

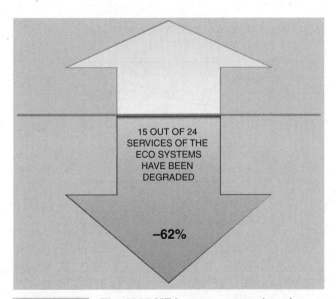

15 OUT OF 24 SERVICES OF THE ECO SYSTEMS HAVE BEEN DEGRADED

−62%

Figure 1.2   The 2005 MEA was a comprehensive analysis by 1360 scientists of 24 benefits or services derived from ecosystems. After four years of consultations and research, the results showed a 62 percent decline in 15 out of 24 services.

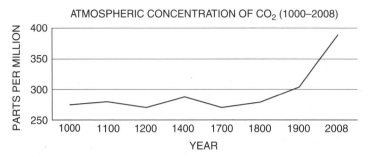

Figure 1.3 Atmospheric CO_2 has increased from a prein-dustrial concentration of about 280 parts per million to about 367 parts per million today. *(Sources: Current and long term historical data compiled by Scripps, Earth Policy Institute, ESRL/NOAA, Worldwatch)*

our means: natural assets and human well-being: statement from the board) show that human activity is putting such strain on natural resources that the ability of the global ecosystem to sustain future generations can no longer be taken for granted. The Ecological Footprint, a conservative measure of natural resource consumption, calculates the total amount of land the world's countries need to produce the resources they use to absorb the waste generated from energy used, and to provide space for infrastructure (WWF 2008). According to this source, man has exceeded the earth's ecological capacity, and we have been living beyond our means since 1987 (see Fig. 1.4).

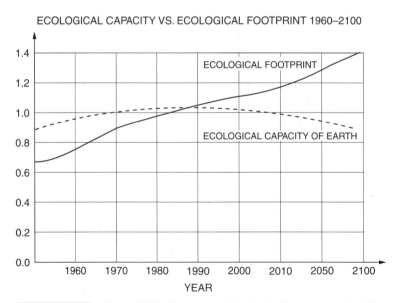

Figure 1.4 Since 1987, Ecological Footprint has exceeded the earth's biocapacity. The United Nations estimates that by 2100 humanity's demand on nature will be more than twice the biosphere's productive capacity. *(Source: WWF, Global Footprint Network, 2008)*

REDUCTION AND FRAGMENTATION OF HABITATS AND LANDSCAPES

The expansion of man-made activities into the natural environment, caused by urbanization, recreation, industrialization, and agriculture have resulted in increasing reduction, disappearance, fragmentation, or isolation of habitats and landscapes. With the extension of cultivation areas from preagricultural times to 1994, the changes are dramatic in terms of loss of forest, woodland, and grassland (see Table 1.1).

As a result, landscape uniformity has been altered, geomorphic processes have been affected, and the quality and the quantity of some natural waters have been changed. The nature of the entire landscape has been transformed by human-induced vegetation changes. (Hannah, Lohse, Hutchinson, Carr and Lankerani 1994; Hannah, Carr and Lankerani 1995; Goudie 2006). The spread of agriculture has transformed land cover at a global scale. Even in the past decade, the cropland areas have quadrupled.

According to the 2005 MEA report, agricultural land has been expanding in about 70 percent of the world's countries, while forest areas are decreasing in two-thirds of those same countries. While there is a slight increase in forest area in the past 30 years in industrial countries, we see a 10 percent decline in developing countries over the same time period (Millennium Ecosystem Assessment 2005). Significant deforestation in tropical forests has been documented for 1990 to 2000. The total loss of natural tropical forests is estimated for this period at 15.2 million hectares per year (FAO 2001) (see Fig. 1.5.).

MASS EXTINCTION OF SPECIES (PLANTS AND ANIMALS)

Although the extinction of plants and animals is a natural part of evolution, the number of species disappearing per year has increased dramatically since the 1800s and directly correlates to population growth and the impact of man-made activities on the environment. The 2008 Living Planet Report, co-prepared by the World Wildlife Fund

TABLE 1.1   ESTIMATED CHANGES IN THE AREAS OF THE MAJOR LAND COVER AREAS BETWEEN PREAGRICULTURAL TIMES TO 1994

LAND COVER	PREAGRICULTURE AREA	PRESENT AREA	PERCENT CHANGE
Total forest	46.8	39.3	−16.0%
Tropical forest	12.8	12.3	−3.9%
Other forest	34	27	−20.6%
Woodland	9.7	7.9	−18.6%
Grassland	34	27.4	−19.4%
Cultivation	1	17.6	1760.0%

Source: Modified from Meyer and Turner, 1994.

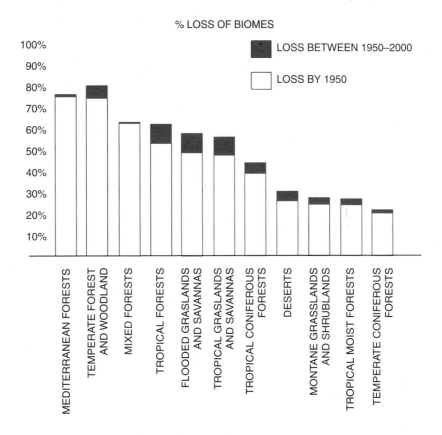

% LOSS OF BIOMES

Figure 1.5   **Conversion of terrestrial biomes.** *(Sources: Millennium Ecosystem Assessment 2005, Ecosystems and Human Well-being: Current State and Trends ; UNEP/GRID-Arendal Maps and Graphics Library (2008))*

(WWF), Global Footprint Network, and the Zoological Society of London (ZSL), states that the living planet index, which charts the populations of species of animals and plants, has declined by a third over the past 35 years (WWF 2008) (see Fig. 1.6).

According to the Threatened Species Red List released by the International Union for the Conservation of Nature and Natural Resources (IUCN) in 2008, 40.1 percent of all species, approximately 19.2 percent of the animals and 20.9 percent of the plants are classified as "threatened" (see Fig. 1.7). These threats are directly linked to the loss of habitats due to destruction, modification, and fragmentation of ecosystems as well as from overuse of chemicals, intensive farming methods, hunting, and general human disturbance. The overall deterioration of global air and water quality, excavations, and deforestation, add to the detrimental influence. Looking solely at species loss resulting from tropical deforestation, extinction rate forecasts climb as high as 75 percent (Ehrlich 1985; Clark, Reading and Clarke 1994; Broswimmer 2002).

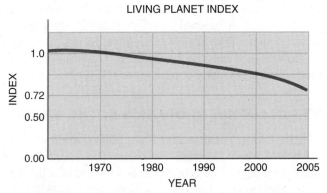

LIVING PLANET INDEX

Figure 1.6 Global Living Planet index shows 28 percent decrease in 4642 populations of 1686 species since 1970. *(Source: Millennium Ecosystem Assessment 2005, Ecosystems and Human Well-being: Synthesis)*

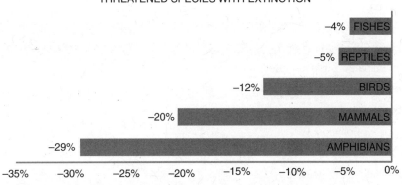

THREATENED SPECIES WITH EXTINCTION

Figure 1.7 The 2008 Red List, prepared by the IUCN World Conservation Congress shows at least 1141 of the 5487 mammals on earth are known to be threatened with extinction. *(Source: IUCN Red List, 2008)*

POLLUTION AND CLIMATE CHANGE

Since Paleolithic times, humans have had some effect on the environment, and pollution has always been a part of human history. As summarized in the Brief History of Green Architecture, chapter 3, the rise and fall of all existing civilizations are connected to environmental pollution and its subsequent consequences. However, any type of pollution, whether of land, water, or air, before the twentieth century was more or less local. But beginning in the early 1900s, especially after World War II, globalization, increasing population, and the use of industrial processes created a new paradigm, in which our modern way of life began to have a much greater collective impact on our surroundings than ever before.

Today, harmful emissions into the air and water from urban, industrial, and agricultural sources affect over a billion people around the world by making resources either unusable or unhealthful. The World Bank estimates that about 20 percent of health concerns in developing countries can be traced to environmental factors (WB 2007). Pimentel et al. study reports the number could be higher (Pimentel, Tort, D'Anna, Krawic, Berger, Rossman, Mugo, Doon, Shriberg, Howard, Lee and Talbot 1998) with 40 percent of deaths resulting from exposure to environmental pollutants and malnutrition (WHO 1992; WHO 1995).

TOP TEN WORST POLLUTION PROBLEMS IN THE WORLD

In a joint report with Green Cross Switzerland in 2008, Blacksmith Institute produced the first list of the "World's Worst Pollution Problems: The Top Ten of the Toxic Twenty," an overview of the range of pollution threats which details the sources and effects of pollution in the most polluted places around the world (see Table 1.2). The 20 major global pollution problems are directly associated with economic and technological factors of the regions. High levels of urbanization, poor or no infrastructure, and lack of formal sector employment, as well as over leveraged governments, present very dangerous conditions for human health, as people turn to informal and often toxic sources of generating income (Blacksmith_Institute 2008).

TABLE 1.2   TOP TEN WORLD'S WORST POLLUTION PROBLEMS
Contaminated Surface Water
Groundwater Contamination
Urban Air Quality
Indoor Air Pollution
Metals Smelting and Processing
Industrial Mining Activities
Radioactive Waste and Uranium Mining
Untreated Sewage
Used Lead Acid Battery Recycling
Artisanal Gold Mining
Sources: Blacksmith_Institute and Green Cross joint report (2008).

Although each one of these issues has a devastating impact on the health of the environment and its people, contaminated surface water, groundwater contamination, urban air quality, and indoor air pollution are directly and/or indirectly related to architecture, and therefore, need to be addressed.

Contaminated surface water now affects one-third of the world, and almost five million annual deaths in the developing nations are due to water-related diseases (Prüss-Üstün, Bos, Gore and Bartram 2008).

Groundwater pollution is another major health issue connected to surface water contamination. Groundwater makes up 97 percent of the world's accessible freshwater reserves and only 0.3 percent of it is useable for drinking (WB 2008). An overwhelming number of people in developing countries rely heavily on groundwater, mostly from shallowly dug wells. These can easily become polluted, primarily through human activities (Blacksmith_Institute 2008).

Air pollution, indoor and outdoor, is another source of health problems. More than half of the world's population relies on unprocessed biomass fuel, such as wood, animal and crop waste, and coal to meet their basic needs. Cooking and heating with such solid fuels without exhaust systems leads to indoor air pollution. Every year this is responsible for the death of 1.6 million people worldwide (WB 2007). Urban areas have an equally dangerous exposure to outdoor air pollution, even though national governments and multilateral development organizations widely recognize the health hazards involved. The World Health Organization estimated in 2007 that 865,000 deaths per year can be directly attributed to outdoor air pollution (WB 2007). As seen in the pollution demographics tables (see Tables 1.3 and 1.4), most studies on the health effects of outdoor air pollution have focused on urban environments of 100,000 plus persons, where the impact is considered to be most severe, since people are routinely exposed to heavy concentrations of airborne pollutants.

However, not only does air pollution harm the environment, it also influences and changes climate. The report of the Earth System Research Laboratory (ESRL) emphasizes the direct impact of air pollution on climate, including greenhouse gases such as carbon dioxide, sulfur, and nitrogen dioxide. Although these gases do not make up a large percentage of earth's atmosphere, even in small quantities they have a profound effect on global climate. Greenhouse gases are mostly responsible for the increase in global surface air temperature of about 0.6°C (1°F) over the past century, and scientists project that much more warming will likely happen during the next century (ESRL 2009).

TABLE 1.3   AIR POLLUTION RATE IN THE WORLD. MANY OF THE WORLD'S MOST POLLUTED CITIES ARE LOCATED IN CHINA AND INDIA, AND OTHER DEVELOPING INDUSTRIALIZED COUNTRIES

CITY	PARTICULATE MATTER ($\mu g/m^3$)	SULFUR DIOXIDE ($\mu g/m^3$)	NITROGEN DIOXIDE ($\mu g/m^3$)
Chongqing, China	123	340	70
Tianjin, China	125	82	50
Shenyang, China	101	99	73
Cairo, Egypt	169	69	..
Kolkata, India	128	49	34
Delhi, India	150	24	41
Lucknow, India	109	26	25
Tehran, Iran	58	209	..
Taiyuan, China	88	211	55
Guiyang China	70	424	53

(Source: Compiled from World Bank Statistics—World Development Indicators (WB 2007))

TABLE 1.4   AIR POLLUTION BY COUNTRY. URBAN POPULATION WEIGHTED AVERAGE PM10 CONCENTRATIONS (MICRO GRAMS PER CUBIC METER) IN RESIDENTIAL AREAS OF CITIES LARGER THAN 100,000

COUNTRY	2000 URBAN POPULATION	1999 PM10 CONCENTRATION
Sudan	5094060	246
Mali	1215170	194
Pakistan	37274140	180
Iraq	14833284	178
Uruguay	1473550	173
Niger	1150996	164
Chad	1430623	161
Egypt, Arab Rep.	21101690	152
Bangladesh	10585200	147
Kuwait	1638081	134

(Source: Compiled and modified from the World Bank Statistics –World Development Indicators (WB 2007), and World Bank Air Pollution in World Cities database (Wheeler, Deichmann et al. 2007))

In 2008, all indicators—atmospheric concentration of CO_2, carbon emissions from burning fossil fuels, and global land-ocean temperature at earth surface—showed steady increases (see Figs. 1.8 and 1.9). The average atmospheric carbon dioxide concentration reached 387 parts per million (ppm), up almost 35 percent since the industrial revolution

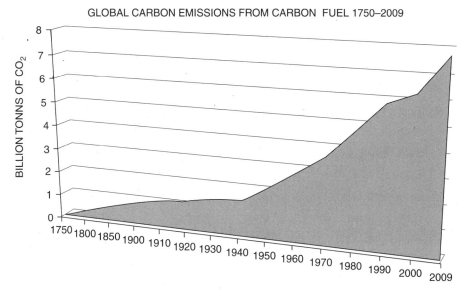

GLOBAL CARBON EMISSIONS FROM CARBON FUEL 1750–2009

Figure 1.8   Global CO_2 emissions have risen sharply since the Industrial Revolution. *(Source: ESRL, 2009)*

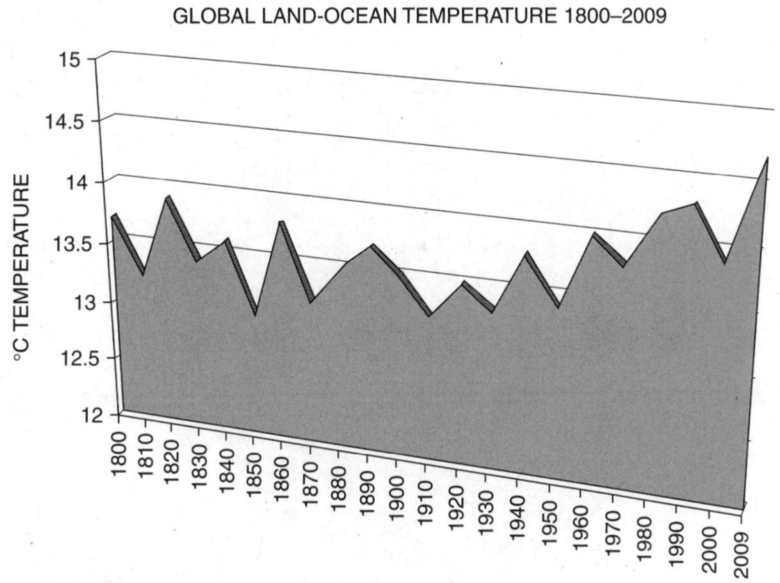

GLOBAL LAND-OCEAN TEMPERATURE 1800–2009

Figure 1.9 **Global land-ocean temperatures are in steady increase since 1960. In 2009, global temperatures increased more than 20 years average.** *(Source: GISS, 2009)*

and the highest for at least the last 650,000 years (ESRL 2009). These recent figures clearly indicate that man-made activities have weakened natural "sinks," such as forests, seas, and soils that absorb carbon.

Architectural Impact on the Natural Environment

Architecture is a complex discipline that incorporates a variety of fields, including design, planning, building construction, landscaping, engineering, and social sciences. Therefore, it is difficult to sort out particular disciplinary areas or to pinpoint specific issues that have an environmental impact. The environmental factors associated with architecture are mostly holistic, far-reaching, and long-lasting, ranging from the resources expended to construct the buildings (technology, materials, energy, water, transportation), the waste produced during construction, the resources used by occupants over the lifetime of the building, and finally the resources and waste involved with demolition and recycling of the structure. Additionally there are the associated problems arising from the wider infrastructural issues including transportation, parking, drainage, and provision of services, all of which make up the urban building scene.

From the environmental viewpoint, buildings account for nearly half of all energy consumption and raw material use around the globe (Fig. 1.10). According to the 2008

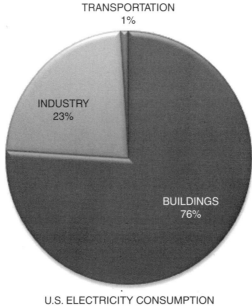

TRANSPORTATION
1%

INDUSTRY
23%

BUILDINGS
76%

U.S. ELECTRICITY CONSUMPTION

(a)

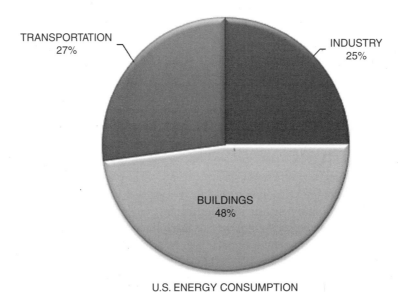

TRANSPORTATION
27%

INDUSTRY
25%

BUILDINGS
48%

U.S. ENERGY CONSUMPTION

(b)

Figure 1.10   Buildings are responsible for (a) 76 percent of the electricity and (b) 48 percent of the energy consumption in the United States. *(Sources: American Institute of Architects (AIA) and Architecture2030 2009)*

Buildings Energy Data book (USDE 2008), commercial and residential buildings account for 39.7 percent of the energy consumed (residential 21.5 percent and commercial 18.2 percent). They are responsible for 76 percent of the electricity used and 15 percent of the total water consumed (Architecture 2030 2009). Similar data was reported by the International Energy Agency (IEA) in 2005, which estimated that buildings accounted for approximately 40 percent of worldwide energy use, equivalent to 2500 Mtoe (million ton oil equivalent) per year (IEA 2005). Studies carried out by the Organization for Economic Co-operation and Development (OECD) indicate that energy consumption by the building sector in OECD countries has continually increased since the 1960s (OECD 2003) and will continue to do so in the coming years, mainly due to construction booms in Asia, the Middle East, and Latin America (UNEP 2006).

At the same time, the building and construction sector takes the largest share of natural resources, both for land use and for materials extraction. Buildings use 50 percent of the world's raw materials—many of which are nonrenewable resources—and they are responsible for 36 percent of all waste generated worldwide (Graham 2002).

Buildings also account for one-third of greenhouse gas emissions (GHG) worldwide, which will increase sharply as construction increases. Estimated at 8.6 billion tons in 2004, building-related GHG emissions could almost double by 2030 to reach 15.6 billion tons under high-growth construction, according to the Intergovernmental Panel on Climate Change (IPCC 2007) (see Fig. 1.11).

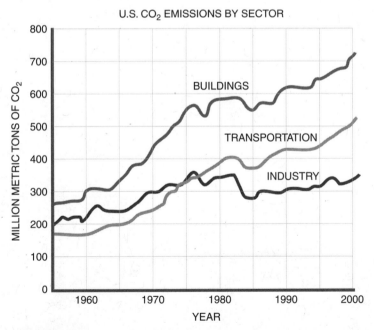

U.S. CO_2 EMISSIONS BY SECTOR

Figure 1.11 In comparison to other sectors, buildings are the largest energy consuming and CO_2 emitting sector.
(Sources: U.S. Energy Information Administration statistics (USDE 2008), Architecture2030)

Although there are various, some even uncontrollable, reasons contributing to this unsettling scenario, three major factors seem to play important roles in environmental impact.

- Overpopulation and resource (un)availability
- Complex construction processes and life span of buildings
- Technologies and materials

OVERPOPULATION AND RESOURCE (UN)AVAILABILITY

A growing population, the emergence of new towns, and the consequent need for housing, infrastructure services, and construction activities mean an increased demand for resources. Over-population especially places competitive stress on basic life-sustaining resources, such as food, shelter, energy, clean water, and clean air (Nielsen 2006). From the beginning of human history to the turn of the nineteenth century, world population grew to a total of one billion people. During the 1800s, population rose at increasingly higher rates, reaching a total of about 1.7 billion people by 1900. According to the U.S. Census Bureau, the world's population in 2008 was 6.7 billion—quadrupled since the turn of the twentieth century (U.S. Census Bureau 2008) (see Figs. 1.12 and 1.13). The

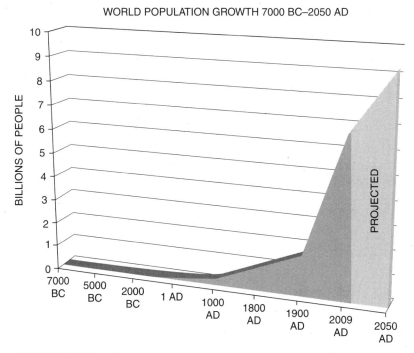

Figure 1.12   World population increased more than 600 percent between 1800 and 2009 and is projected to increase by an additional 37 percent to 9.2 billion by 2050. *(Sources: UN Population Division, 2006; U.S. Census Bureau, 2008)*

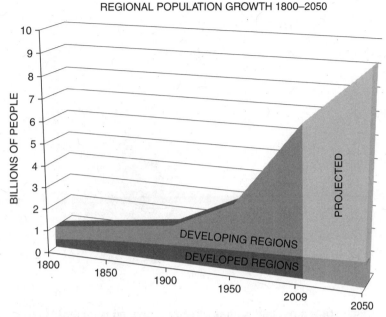

REGIONAL POPULATION GROWTH 1800–2050

Figure 1.13   **Population ratio between developed and developing regions is projected to reach 1/8 by 2050.** *(Sources: UN Population Division, 2006; U.S. Census Bureau, 2008)*

United Nations projects that the world population will increase by 2.5 billion over the next 41 years, from the current 6.7 billion to 9.2 billion by 2050. The projected increase alone is equivalent to the total world population in 1950. Less developed countries, whose population is projected to rise from 5.4 billion in 2007 to 7.9 billion in 2050 will probably see most of the increase (U.N. Population Division 2006).

In addition to depletion of natural resources, pollution, poverty, and malnutrition, one of the major problems associated with or exacerbated by overpopulation is the high-density urban settlements. The United Nations projects that by 2010, for the first time, the urban population of the earth will outnumber the rural. In 1800 only 3 percent of the world's population lived in cities. In 1950, there were only 86 cities with a population of more than one million; today, there are 400, and by 2015, there will be more than 550 (Davis 2006; U.N. Population Division 2006). This rapid urbanization means that over the next 30 years developing countries are predicted to triple their population size and account for 80 percent of the world's urban population (see Fig. 1.14).

This being the case, the majority of this population, 60 percent, will live in slums. Today over one billion people already live in slums, with this figure rising by 25 millions a year, and only 1 percent of housing and urban aid is given to urban slums (see Table 1.5). Although Asia presently has over half the world's slum population (581 million), Africa is expected to exceed this percentage by 2020 (Davis 2006; U.N. Population Division 2006).

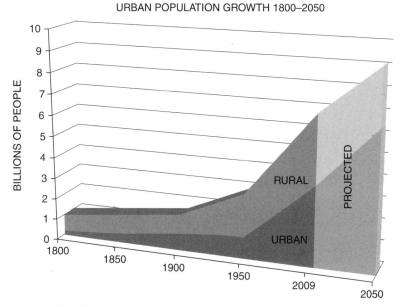

URBAN POPULATION GROWTH 1800–2050

Figure 1.14   **Urban population has already exceeded rural and is projected to increase 40 percent by 2050** *(Sources: UN Population Division, 2006; U.S. Census Bureau, 2008)*

As the world population continues to grow, our natural resources are under increasing pressure. This threatens our public health and social and economic development. This is especially true in developing countries, where people consume resources much faster than they can be renewed. According to a 2000 report published by the Johns Hopkins Population Information Program (Hinrichsen and Robey 2000), living conditions are worsening in every environmental sector—public health, food supply, fresh water, coastlines and oceans, biodiversity, and climate change.

- *Public health.* Contaminated, unclean water (along with poor sanitation and air pollution) kills over 15 million people each year, many of whom live in developing countries. Public health is also seriously threatened by heavy metals and other contaminants polluting our environment.
- *Food supply.* According to a United Nations Food and Agriculture Organization's study, in over 60 percent of developing countries the population has been growing faster than its food supply. This disparity has degraded over two billion hectares of arable land.
- *Fresh water.* The supply of fresh water is limited, but demand is increasing as the population grows and use-per-capita rises. By 2025, the world population is projected to be eight billion, forcing people to cope with fresh water shortages. Soaring demand is placing a tremendous amount of pressure on the world's water resources, resulting in disappearing rivers and dropping water tables. According to the 2007 World Bank report, nearly 70 percent of global water withdrawals from rivers,

TABLE 1.5   POPULATION IN MEGACITIES IN THE DEVELOPING WORLD IS EXPANDING*

CITY	1950	2004	2009
Mexico City	2.9	22.1	22.9
Seoul	1.0	21.9	23.9
Sao Paulo	2.4	19.9	21.1
Mumbai (Bombay)	2.9	19.1	22.3
Jakarta	1.5	16.0	15.1
Dhaka	0.4	15.9	13.1
Calcutta	4.4	15.1	16.0
Cairo	2.4	15.1	14.8
Manila	1.5	14.3	19.2
Karachi	1.0	13.5	15.7
Lagos	0.3	13.4	11.4
Shanghai	5.3	13.2	17.9
Buenos Aires	4.6	12.6	14.1
Rio de Janeiro	3.0	11.9	12.5
Tehran	1.0	11.5	12.5
Istanbul	1.1	11.1	12.5
Beijing	3.9	10.8	13.2
Bangkok	1.4	9.1	9.2
Gauteng	1.2	9.0	9.6
Kinshasa	0.2	8.9	8.6

*Developing world megacities (population in millions).

(Sources: UN-HABITAT Urban Info (2009); Davis 2006; Brinkhoff 2009)

lakes, and aquifers are used for irrigation, while industry and households account for 25 and 10 percent, respectively. The 2000 U.S. Geological Survey (USGS) study finds that the total water usage in buildings is 9.4 percent, and residential use is almost three times more than the commercial buildings (see Table 1.6).

■ *Coastlines and oceans.* More than 50 percent of coastal ecosystems are under pressure by high-density populations and urban dwellings. For instance, coastlines and oceans are becoming increasingly more polluted. In addition, overfishing is reducing the number of fish catches within these areas.

TABLE 1.6   ACCORDING TO THE 2000 USGS SURVEY, BUILDINGS USE 9.4 PERCENT OF THE FRESH WATER*

YEAR	ALL BUILDINGS	% OF TOTAL WATER USE	RESIDENTIAL	% OF TOTAL WATER USE	COMMERCIAL	% OF TOTAL WATER USE
1985	31.26	7.8%	24.32	6.1%	6.94	1.7%
1990	33.58	8.2%	25.29	6.2%	8.29	2.0%
1995	35.67	8.9%	26.09	6.5%	9.58	2.4%
2000	38.34	9.4%	28.03	6.9%	10.31	2.5%

*Total use of water by buildings (billion gallons per day).

(Sources: Hutson, Barber, Kenny, Linsey, Lumia, and Maupin 2004; USDE 2008)

■ *Biodiversity*. The earth's biodiversity is in imminent danger; human activities are forcing numerous plants and animals into extinction. In fact, it is estimated that 65 percent of all species are declining in numbers.

■ *Global climate change*. The planet's surface is warming at an alarming rate, in large part because of the building industry's emission of greenhouse gases. If global temperatures continue to increase as projected, sea levels would rise by several feet. The result would be widespread additional climate change, followed by flooding and drought. Agriculture production would be severely disrupted and/or curtailed, placing additional pressure on the world's food supply.

The depletion of resources is an inevitable result of overpopulation and its subsequent high-density housing needs. There is an unsustainable demand/resource ratio on the earth's natural resources, materials, and energy. Each year, three billion tons of raw materials—about 40 to 50 percent of the total amount consumed by the global economy—are used in the manufacturing of building products and their components (Roodman and Lenssen1995; Anink et al. 1996). Furthermore, the global consumption of key raw materials (such as steel, aluminum, plastic, and cement) is rising fast. For example, over a 20-year period ending in 1994, the world consumption of cement increased by 77 percent and plastic consumption increased by 200 percent, while the world population increased by 40 percent (University of Minnesota, 1999).

COMPLEX BUILDING CONSTRUCTION PROCESS AND LIFE CYCLE OF BUILDINGS

Contrary to common belief, the term architecture does not refer simply to the end product. It involves a long process with various stages, including: mining for resources, manufacturing, transportation of the materials and technology, building, maintenance, demolition, and recycling (see Fig. 1.15). Every stage has its own activities, some more complex than others, with various levels of environmental impacts.

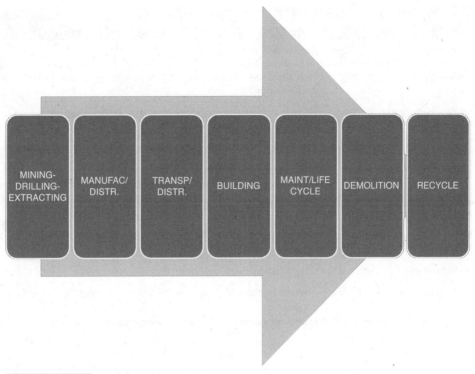

Figure 1.15   **The construction process.**

The construction process starts with mining, drilling, and extracting the materials. Excavating mineral ores for construction requires the stripping of topsoil and rocks. This separation requires energy primarily from fossil-based resources, an activity that can cause much greater harm than the benefits of excavation (Jackson 1996). The more complex the process to obtain the product, the greater the amount of energy consumed. This stage of the construction process is especially harmful for the relevant ecosystem, as it initiates the destruction of existing settlements, plant and animal habitats, land erosion, water pollution, and deforestation.

Manufacturing of construction materials requires more energy, produces waste, and pollutes natural resources. The distribution and transportation of construction materials and technology also impact the environment by using additional energy to transport them from the manufacturing point to the point of assembly and building. Construction activities necessary to complete a building contribute to air pollution, including: land clearing, engine operations, demolition, burning, and working with toxic materials. In fact, all construction sites generate high levels of pollutants, mostly from concrete, cement, wood, stone, and silica. Construction dust, in particular, is a serious issue. Although it is invisible to the naked eye, the dust penetrates deeply into the lungs and causes a wide range of health problems, including respiratory illness, asthma, bronchitis, and cancer (see Table 1.7).

TABLE 1.7   ENVIRONMENTAL IMPACTS OF ARCHITECTURAL PROCESS

ACTIVITY	ENVIRONMENTAL IMPACTS
Mining/drilling/extracting	Deforestation; destruction of plant and animal habitat; existing settlements; land erosion; water pollution
Manufacturing/assembly	Energy consumption (impacts of producing energy); waste generation
Transportation/distribution	Energy consumption, CO_2 emission; resource use (packaging)
Building	CO_2 emission; pollution and radiation from the materials and technologies (exposed to chemical and climatic activities); pressure and damage
Maintenance/life cycle	Energy consumption, CO_2 emission; resource use and replacement; wear and tear; chemical contamination (material loss—from roofs, pipes); water pollution
Demolition	Chemical contamination; toxicity; environmental poisons
Recycle/waste	Landfill decomposition; groundwater contamination; methane gas production

As for the environmental impact and energy consumption, the most harmful stage of the construction process is the maintenance and life cycle of the buildings. The life cycle of a building is a long-lasting process after construction, which includes the performance, durability and maintenance, energy use and consumption, water and waste management, environmental human health systems, occupant well-being, renovation, recycle content, resource limitation, and the life span of the building (see Fig. 1.16).

Because of the complexity and the life span of buildings, there is a consistent flow of materials and technologies during the life cycle (Vogtlander 2001). The life span of buildings is an important factor in this cycle and contributes directly on the buildings' environmental impact. Depending on the category and building type, an average building life is approximately 35 to 50 years (Addington and Schodek 2005) but this number changes based on the category and purpose of the building (see Table 1.8). The actual lives of buildings are affected by various external factors outside the control of the original intentions (Fernandez 2005), and depend on a vast amount of natural resources, including land, energy, and water. The majority of the environmental impact and energy consumption takes place during this stage.

Building demolitions are often impacted by environmental concerns, such as excessive CO_2 emission, toxic materials, solid waste, nonrenewable landfill materials, and health issues.

Most of the construction and demolition materials (e.g., lead-based paints, asbestos, mold, wastes containing mercury, fluorescent bulbs, batteries) pose serious environmental and health problems (Roodman and Lenssen 1995; Berge 2000). For both regulatory and economic reasons, recyclables (such as concrete, lumber, and brick) are

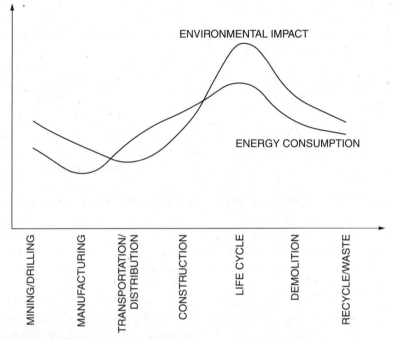

Figure 1.16 The majority of the environmental impact and energy consumption takes place during the life cycle stage.

TABLE 1.8 CATEGORY OF DESIGN SERVICE LIFE FOR BUILDINGS		
CATEGORY	**DESIGN SERVICE LIFE**	**EXAMPLES**
Temporary offices,	Up to 10 years	Nonpermanent construction buildings; sales bunkhouses temporary exhibition buildings
Medium life	25–49 years	Most industrial buildings; most parking structures
Long life	50–99 years	Most residential, commercial, and office buildings; health and educational buildings; parking structures below buildings designed for long life category
Permanent	Minimum period, 100 years	Monumental buildings (e.g., national museums, art galleries, archives); heritage buildings
Source: CSA (R2001) Guideline on Durability in Buildings		

typically separated from other solid waste. As their processing and disposal procedures are minimal and inexpensive, most of these materials are salvaged and reused. Other recyclables (such as steel, aluminum, copper, and glass) possess significant economic value to specialty recycling and salvage facilities. Hazardous waste must be disposed of in a separate landfill at a very high cost. This includes materials with high levels of fossil fuel, chromium, or lead-based contaminants (Gockel 1994).

TECHNOLOGIES AND MATERIALS

Materials, technology, and architecture have had a strong relationship from the very early beginnings of construction. This relationship is almost inseparable and is one of subordination. The material is merely the means of completing a building, and the act of building requires a technology. The invention and use of these elements changed our built environment, but we are only now learning that this connection has had ecological consequences.

The technologies and materials for construction and then operation of buildings have a disproportionate impact on the natural environment when compared to its role in the economy. Although the construction sector represents only about 8 percent of gross domestic product (GDP) in the United States, it consumes 40 percent of all extracted materials, produces one-third of the total landfill waste stream, and accounts for 39 percent of national energy consumption for its operation (Roodman and Lenssen 1995). Raw materials for the building sector are extracted, processed, transported, fashioned in the construction phase, demolished, and recycled. As stated earlier, all these stages imply a number of environmental impacts. In particular, the building industry is a heavy consumer of materials with high-embodied energy content, such as aluminum, cement, and steel, the production of which usually depends on the use of fossil fuels, resulting in CO_2 emissions (UNEP 2006). Lightweight construction materials, such as timber frames, usually have lower-embodied energy in comparison, but because of massive harvesting of this material, 20 percent of the earth's forests have disappeared. Around the world, mining of copper, bauxite, and iron ore resources for building materials continues, pouring large quantities of pollutants into nearby air and water. All these trends are accelerating, and the damage they have done and may do is often irreversible (Roodman and Lenssen 1995).

Throughout the entire construction process, fossil fuel–based energy consumption is high. For example, the high temperatures necessary to produce steel, glass, and brick require great amounts of fossil fuels. Transporting materials to a building site burns yet more fossil fuels. It is not surprising then, that the amount of carbon dioxide in our atmosphere has risen approximately 30 percent since 1900, one-quarter of which comes from fossil fuel combustion used to provide energy for buildings (Roodman and Lenssen 1995).

As mentioned earlier, a completed building does not mean less consumption of fossil fuel and energy, but rather more. According to the 1992 OECD report, energy use in buildings from 1971 to 1992 was at an average 2 percent annually. In 1992, total energy use in buildings had risen to 34 percent. This included 25 percent from fossil fuels, 44 percent from hydropower, and 50 percent from nuclear power. Adding in the fuels and power used in construction, buildings consume at least 40 percent of the world's energy. They thus account for about a third of the emissions of heat-trapping carbon dioxide from fossil fuel burning, and two-fifth of acid-rain-causing sulfur dioxide and nitrogen oxides. Buildings also contribute to other side effects of energy use—oil spills, nuclear waste generation, river damming, toxic runoff from coal mines, and mercury emissions from coal burning (OECD 1992; Roodman and Lenssen 1995).

A survey of water use in buildings tells a similar story. An increase in water use is lowering water tables and necessitating large projects that siphon water supply away from agriculture. In addition, electric power plants use water as a coolant, which then drains into rivers, carrying thermal and chemical pollution. These two uses contribute about equally to buildings' one-sixth share of global water withdrawals (Roodman and Lenssen 1995; Gleick, Cooley, Katz and Lee 2007).

Steel production can be highly polluting; iron mining produces tailings that can leach heavy metals into nearby streams; and open-hearth steel making can emit lead and other poisonous heavy metals. According to 1992 U.K. data, the use of materials such as steel, copper, aluminum, and concrete makes each square meter of floor space in a large office building 2 to 4 times as energy intensive—and therefore approximately 2 to 4 times as pollution intensive as a house (Roodman and Lenssen 1995). Another concern with materials is their potential impact on indoor air quality. Most bonding and drying agents in carpets, veneers, particle board, plywood, and petroleum-based paints emit health-threatening volatile organic compounds (Roodman and Lenssen 1995). Finally, many modern buildings also create dangerous indoor environments for their inhabitants. For example, "sick building syndrome" is reported to occur in 30 percent of new or renovated buildings worldwide (Roodman and Lenssen 1995).

Responsibility of Architecture

Due to its role, volume, and impact, it is evident that architecture has a direct responsibility to the immediate (city) ecosystem. As Graham (Graham 2002) points out, every architectural artifact, regardless of its size—big or small (1) connects to the earth; (2) depends on nature for resources; (3) causes environmental change; and (4) affects both human and nonhuman life. Since part of the problem is architectural, so should be the solution, such as designing based on sustainable and ecological principles; developing and using advanced green technologies and materials; and promoting and demanding high-performance buildings. Some of these issues, such as ecological design have been around for decades. Others have been proposed and promoted, but sporadically rather than consistently.

Architecture's main responsibility is not to pick and choose the "best" solution but to incorporate all options that might generate workable solutions. There is no single formula of what and how much to use. Clearly, there is an urgent need of a new way of thinking and designing. In order to fully address its responsibilities, architecture should abandon old methods, technologies, and materials and push for a new paradigm shift. The design objectives should be based on sustainable, ecological, and performance criteria rather than trends and aesthetics; be environmentally conscious rather than market-driven; and be inherently resourceful rather than globally destructive. Briefly described, the responsibility of green architecture includes:

■ *Smaller buildings*. Unlike Mc-Mansions, smaller buildings are economically feasible, efficient, and require low maintenance. Because of their compact size, smaller buildings use less material, need less energy, and produce less waste. Architects should focus on small, yet functional and ecologically sensitive buildings by conserving space and preserving the environment.

■ *Sustainable materials and technologies.* Architects should focus on using durable, low-maintenance, recyclable, and economical materials and technologies. Constant breakdowns, wear-and-tear, and replacement of materials and technologies will make buildings unsustainable. Using abundant, local elements—if possible—with little to no transportation costs is highly preferable. Architects should also consider elements that are easily dismantled and reused or recycled at the end. They can be salvaged, refurbished, or remanufactured, including saving materials and technologies from disposal and renovating, repairing, restoring, or generally improving the appearance, performance, quality, functionality, or value.

■ *Ecological materials and technologies.* Materials and technologies should consist of low-emission, nonpollutant elements with low manufacturing impacts. Ecological materials should facilitate a reduction in polluting emissions from building maintenance and should not be made from toxic chemicals. Architects should focus on clean burning technologies by excluding the components such as substances that deplete stratospheric ozone and associated with ecological damage and health risks, including mercury and halogenated compounds, and HCFCs (hydrochlorofluorocarbons). Additional ecological technologies such as stormwater and wastewater systems that reduce surface water and groundwater pollution should be incorporated.

■ *Sustainable resources.* Buildings should rely on sustainable resources, such as energy and water, focusing on supplying their own gray water and power. Such buildings may operate entirely off the power grid, or they may be able to feed excess energy back into the grid. Solar, thermal, and wind—if available—powers are the usual alternatives. Buildings should also consider the proximity to and from water resources, supplies, and existing waste management systems. Architects should also consider the climatic conditions for their favor and benefit from them, such as sun, wind, and water. Residential and daylight-needing buildings should not be designed in sun-trapping/blocking areas (i.e., in between buildings, etc.). The buildings should be accessible to public transportation (and bicycle paths) to reduce private vehicle use, to save energy, and to reduce air pollution.

■ *Sustainable environments.* One of the main responsibilities of any architect is to create sustainable environments that are protective, healthy, habitable, and promote social and institutional networks. Buildings should provide protective environments where the occupants feel safe and secure against the various elements such as natural causes, built environments, and people. Building should also provide healthy and habitable environments for people; designed to maximize productivity by minimizing operator fatigue and discomfort; and should be free from physical and psychological effects of buildings such as sick building syndrome.

■ *Resource ecology.* By taking ecological issues into account, architects should design and construct buildings in the right places and in the right way, for the benefit of both the occupants and the ecological resources. The reduction of the natural resource consumption should be targeted right from the start, at the design stage. The calculation and control activities should focus on the building's natural resource use, such as water, energy, landscape, and waste management. Soil type and groundwater conditions must be taken into consideration before the building is designed and constructed. The type and stability of soil should be taken seriously, not only because of the building damage but also potential problems to the soil ecology such as erosion, pollution, sedimentation, and various forms of soil degradation.

■ *Environmental ecology.* One of the main responsibilities of architects is to respect the ecology of the environment, and to design the buildings in a way not to pollute the environment and harm the ecosystem. Faulty and poorly designed and/or installed building infrastructure systems, such as inadequate gray water and sewage pipes, stormwater management, and drainage systems, can contribute to drainage, flooding, and soil and groundwater pollution. Architects should make sure to provide proper drainage systems which collect runoff from impervious surfaces (e.g., roofs and roads) to ensure that water is efficiently conveyed to waterways through pipe networks. Designs should promote minimizing water usage and providing water-efficient landscaping. The materials, technologies, and the type of energy used in the buildings should be selected from nonpollutant elements, such as alternative energy resources and low-VOC building products.

Architects should also implement global stewardship principles by acting locally and thinking globally. Use local resources as much as possible by reducing the embodied energy of the building products, and by considering global ecological consequences of their actions. Land selection, biodiversification, and building orientation should be integrated into the design before the building is constructed. Avoid changing the ecosystems for the sake of building landscape and/or orientation, such as cutting off plants or creating artificial ecosystems, which might contribute to erosion and flooding. Instead, buildings should contribute to the environment by absorbing sun rays and stabilizing the soil.

■ *High-performance materials and technologies.* The materials and technologies used in buildings should be efficient, effective, and productive. The material efficiency can be achieved by using recycled elements with minimal waste or adding engineered components, such as engineered lumber and I-joists. Technological efficiency should apply to the entire building cycle, including water and energy efficiency. These elements should also be effective by producing desired results and productive such as changing and storing the energy and water.

■ *Resource performance.* A building's resource performance is determined by the contribution to the resources of the location. Buildings should perform as economic, ecological, and environmental contributors by various different ways. The location and function of the building should contribute to the economic viability in the area by creating jobs, enhancing property values, and bringing other businesses into the area. The material and technological elements of the building should also be used in a way to reduce the environmental impact of the building such as absorbing sun rays and CO_2 emission from the atmosphere.

■ *Environmental performance.* Buildings should be physically, functionally, and socially adaptable to the environment and perform according to environmental changes. Changes in climate, social patterns, or trends should not end the building life cycle but give birth for different uses for the building. The functional and environmental quality of the building should also be considered as the main design objective.

DEFINITIONS AND

OPERATIONALIZATIONS OF

GREEN ARCHITECTURE

The term "green" is of one the most widely used but poorly defined terms in architecture today. While the terms "green," "sustainable," and "ecological" are often used interchangeably to describe environmentally responsive architecture, in reality each term has its own history and sociopolitical connotations, as well as its own architectural definition, use, and operation. Green architecture is an umbrella term, which involves a combination of values—environmental, social, political, and technological—and thus seeks to reduce the negative environmental impact of buildings by increasing efficiency and moderation in the utilization of building materials, energy, and development space. Therefore, the definitions of the following terms are essential to understand and define green architecture.

Sustainability

The term "sustain," derived from the Latin "sustenere," means to keep in existence, to be capable of being maintained in a certain state or condition (Lawrence 2006). Because of the ecological roots of the term, it is mostly used to address environmental and climatic concerns. However, sustainability was first introduced as a global socioeconomic concept during the 1970s and defined later in "Our Common Future" by Brundtland Commission as "meeting the needs of the present without compromising the ability of future generations to meet their own needs" (UN 1987). Therefore, depending on the context, there are different approaches and definitions of sustainability. For example, in an ecological context, sustainability is defined as the ability of an ecosystem to maintain ecological processes, functions, biodiversity, and productivity into the future (Naiman, Bilby and Kantor 1998; Waltner-Toews, Kay and Lister 2008).

In an architectural context, sustainability is defined as a term that describes economically affordable, environmentally healthy, and technologically efficient and high-performance buildings (Edwards 2005; Steele 2005; Sassi 2006; Smith 2006; Steinfeld 2006; Williams 2007; Newman and Jennings 2008; Vallero and Brasier 2008). Besides architecture and ecology there are other definitions of sustainability in other fields: technological (Dorf 2001; Anastas and Zimmerman 2003; Teich 2003; Olson and Rejeski 2004); material (Spiegel and Meadows 2006; Kibert 2007), economic (Brown 2001; Lopez and Toman 2006), and behavioral (Ehrenfeld 2008; Murphy 2008).

Sustainable Architecture

In architecture, sustainability is used as a general term to describe technologically, materially, ecologically, and environmentally stable building design. Within the context of sustainable architecture, stability is established via three major components:

1 Technological and material sustainability (elements)
2 Resource sustainability
3 Environmental sustainability

Creating sustainable buildings requires that one consider the sustainability of their technological and material elements, resources, and environment. An element's sustainability is measured by its durability, maintenance level, and recyclability. Economic issues related to its construction, profitability, and building stock value should also be considered. Resource sustainability can be measured based on its site condition, cost-effectiveness of the operational and life cycle of the building, accessibility, and favorable natural forces. Finally, creating healthy, habitable, and safe environments with social and institutional capacity should be the primary focus for environmental sustainability. The architect's challenge, therefore, is to find a balance among technological and materials considerations, resource availability, and environmental sustainability (see Table 2.1).

TABLE 2.1 COMPONENTS FOR SUSTAINABLE ARCHITECTURE		
ELEMENTS	**RESOURCES**	**ENVIRONMENT**
Durable	On-site conditions	Healthy
Economical	Cost-effective (operational/life cycle)	Habitable
Low-maintenance	Accessibility	Social/institutional capacity
Recyclable	Natural forces (favorable)	Safety and security (protective)

SUSTAINABLE ELEMENTS

The primary elements of any building are the technologies and materials. These elements play an important role on the sustainability of a building. Their embedded sustainable properties and intrinsic values make up the elements of sustainable architecture.

Durability   Materials and technologies should be strong, resilient, stable, and long lasting. Freedom from constant breakdown, wear and tear, and replacement help the buildings to be sustainable. Green buildings should use durable materials that resist decay, wear, and mold with a high level of tolerance, and technologies that can be used over a relatively long period without being depleted and/or changed.

Economical   Elements should be cost-effective and economical. Abundant local elements with little or no transportation costs should be used if possible. Consider higher efficiency, durability rate, and low-maintenance records when making choices. The economics of technologies should be calculated based on their performance and output ratio, not on the nominal purchasing value. For example, solar panels cost more at the outset but are highly economical, in terms of energy costs, throughout the life cycle of the building.

Low-maintenance   Sustainable elements should be selected from materials and technologies that do not require constant servicing and/or replacement due to wear and tear, deterioration, or decline in performance. Materials should be self-sufficient and carefree. For example, preinsulated siding materials, which are composed of siding adhered to a layer of insulated foam board with special flexible adhesives, make the material solid, rigid, impact resistant, and quite efficient, with much less maintenance because of its compact composite nature. Low-maintenance technologies should be ecofriendly, renewable, and self-sufficient. Advances in energy harvesting technologies, such as thermoelectric, piezoelectric, and others can replace the rechargeable-battery systems as self-sufficient devices. By replacing batteries, these technologies eliminate toxic waste from disposable batteries, reduce power consumption and environmental waste, and require much less maintenance with a very low impact.

Recyclability   Elements should be capable of being easily dismantled, reused, or recycled. Consider whether they can be salvaged, refurbished, or remanufactured, including saving technologies from disposal by renovating, repairing, or restoring. Elements from salvaged products with pre- and postconsumer recycled content, and products made from agricultural waste should be selected.

SUSTAINABLE RESOURCES

Resource sustainability requires four important principles: economy/cost-effectiveness, on-site conditions, accessibility, and (favorable) natural forces.

Economy/cost-effectiveness   In order for a building to be economical and cost-effective, the land, utilities, and operational costs should be affordable. Green construction costs should be comparable, if not lower, to traditional construction materials and labor. Overheads should be estimated to be paid back in less than two years. Green building operating expenses should be significantly lower. Energy savings and water reduction technologies should yield a considerable savings, such as 30 percent or more. Waste generation should be minimized in order to reduce the repairs, maintenance, and landfill tipping fees. In order to reduce operational expenses, indoor air quality should be optimal, which in turn can lead to reduced owner liability by reducing insurance premiums over the life of the building. Moreover, green buildings qualify for subsidies and tax credits for the cost of green technologies and materials. An increasing number of states have started to offer tax credits and additional financing for as much as 50 percent of the cost for green technologies and materials.

On-site conditions   On-site conditions, including location and functional qualities, should be favorable for a green building. These include proximity to the necessary resources and utility lines, such as water supply, gas mains, electrical poles, and waste management systems.

Accessibility   Accessibility to transportation should be included in green building design. Close-proximity, high-density mixed developments should be considered to minimize walking distance, to save energy, and to reduce cost. Easy access to public transportation: bus or metro should be provided. Urban refill and reduced site disturbance should also be considered.

Natural forces   Buildings should be designed based on consideration of existing natural forces and should be built to benefit from them. For example, in cold climates, buildings could be designed in close proximity, such as row-housing units, where common side walls reduce heat loss. The orientation of the building should be selected for proper wind protection (or wind-cooling) and sun exposure. Residential and other building types that require daylight (i.e., hospitals) should be designed carefully so as not to lose solar access. Also important are natural ventilation potential and good ambient air quality. Sensitivity to water quality and runoff minimization issues need to be considered.

SUSTAINABLE ENVIRONMENTS

Green buildings should provide healthy, habitable environments for its occupants. Buildings with comfortable layouts should be structurally sound and use nontoxic materials and technologies with adequate ventilation and proper filtration. Green buildings should also provide protection from natural, social, and building-related events, and provide an environment for social and institutional connections to occur.

Healthy   Green buildings should provide healthy environments for occupants that reduce the risk of various airborne diseases, such as asthma or allergies, to prevent sick building symptoms. Zero- or low-emission building products should be used to improve indoor air quality. Architects should avoid using building materials and cleaning/maintenance products that emit toxic gases, such as volatile organic compounds (VOC) and formaldehyde, which can have a detrimental impact on occupants' health and productivity. Adequate ventilation and a high-efficiency, in-duct filtration system should be provided. Green buildings should offer specific solutions to prevent indoor microbial contamination by: selecting materials resistant to microbial growth; providing effective drainage from the roof and surrounding landscape; installing adequate ventilation in bathrooms; allowing proper drainage of air-conditioning coils, and designing building systems to control humidity.

Habitable   Green buildings should contain at least minimum habitable requirements for buildings, which include physical safety of the building such as load, separation and classification of the occupants, openings, fire protection, construction type, and adequate structural systems. In addition to local codes, the building's structural, construction, electrical, and plumbing codes should be in line with international building codes. In a business setting, buildings should maximize productivity by minimizing operator fatigue and discomfort.

Social/institutional capacity   On the nontechnical level, sustainable environments are responsive to the social networks, activities, and events of the occupants. Green building design should consider prevalent social patterns, attitudes, and networks to create effective solutions for their inhabitants. Buildings should enhance a sense of community and social opportunities, such as providing inviting open spaces and public environments (Lang, Burnette, Walter and Vachon 1974; Lang 1994). Green buildings should consider incorporating the nature and the interrelationship of organizational design and spatial arrangements, the nature of communities and neighborhoods, and the relationship between family organizations and neighborhood layouts.

Safety and security   Green buildings should be designed to provide physical protection to the occupants; for instance, by preventing accidents through the choice of building materials and technologies, or supporting the building to withstand natural disasters like earthquakes, high winds, and floods. Buildings should avoid floodplains, or at least raise ground floors above flood levels. A sufficient and simple means of egress is essential to ensure a safe exit in case of fire. Green buildings should also provide a sense of privacy, clear boundaries, users' control, and personalization for occupants, with clear territorial demarcations. According to an empirical study done by Oscar Newman, people own and take control of an area where territorial demarcations are clear. Newman found that crime rates were lower in areas where (1) a clear set of territorial markers differentiate public spaces from semipublic and private spaces; (2) no undefined open spaces exist; (3) opportunities exist for natural surveillance such as watching other people as part of everyday life through the placement of halls, windows, and seating where people gather and overlook other areas; and (4) the use of building and landscaping forms and materials that communicate a positive, defensible image of the residents to outsiders (Newman 1972; Lang, Burnette, Walter and Vachon 1974; Lang 1994).

Ecology

The term, "ecology" was first coined by Ernst Haeckel (1834–1919) in 1866 as *Okologie*, from Greek *Oikos* and came into English in 1873. Haeckel first defined the term as "the comprehensive science of the relationship of the organism to the environment" (Frodin 2001). As a term, "ecology" has Greek roots, reflecting an early concern with humanity. *Oikos* means home or house in Greek and by extension it means the whole inhabited earth, the *oikoumeme*, the home of all mankind. Logos meaning reason or study, is a common suffix applied to many of the sciences, indicating the human mind at work on a given subject (Hughes 1975).

In general, ecology studies the interactions between organisms and their environments. The environment includes physical properties, which consist of biotic and abiotic factors as explained in Chapter 1. Therefore, ecology is a broad discipline comprising many subdisciplines, such as "ecophysiology," which examines how the physiological functions of organisms influence the way they interact with the environment (Koch and Mooney 1995; Calow 2008); "ecosystem ecology," which studies the flow of energy and matter through the biotic and abiotic components of ecosystems (Odum 1971; Golley 1996; Chapin, Mooney and Chapin 2004; Pastor 2008); "evolutionary ecology," which studies ecology in a way that explicitly considers the evolutionary histories of species and their interactions (Pianka 1999; Fox, Roff and Fairbairn 2001); "behavioral ecology," which examines the roles of behavior in enabling an animal to adapt to its environment (Krebs and Davies 1997; Caro 1998; Dugatkin 2001); and "architectural ecology," which examines the relationships and connections between buildings, occupants, surrounding habitat, cities, and the global ecosystem (Beatley and Manning 1997; Wines 2000; Graham 2002; Broadbent and Brebbia 2006; Newman and Jennings 2008).

Ecological Architecture

Ecological architecture is mainly concerned with how ecological properties impact the building, its occupants, and the environment. The term is generally used as a framework to describe multilevel ecological building design and its balance with nature. This balance is established via three major components (see Table 2.2):

1 Ecological elements (technological and material)
2 Resource ecology
3 Environmental ecology

ECOLOGICAL ELEMENTS

Ecological elements should be selected from natural or minimally-processed earth resources. They should be biodegradable, renewable, and clean elements with low-embodied energy.

TABLE 2.2 COMPONENTS FOR ECOLOGICAL ARCHITECTURE		
ELEMENTS	**RESOURCES**	**ENVIRONMENT**
Clean (nonpollutant/ low-emission)	Resource share	Pollution (air/water/land)
Earth resources	Soil/landscape	Global stewardship
Biodegradable	Site selection	Biodiverse
Low-embodied energy	Water resources and use	Land use
Renewable	Waste management (low solid waste)	

Clean (nonpollutant/low-emission)   Materials and technologies should consist of low-emission, nonpollutant elements with low manufacturing impacts. Ecological materials should facilitate a reduction in polluting emissions from building maintenance and should not be made from toxic chemicals. These materials clearly exclude the components (1) with substances that deplete stratospheric ozone and (2) are associated with ecological damage or health risks, including mercury and halogenated compounds. Ecological elements should avoid ozone-depleting substances (ODS), such as hydrochlorofluorocarbon (HCFC) and hydrobromofluorocarbon (HBFC).

Green buildings should incorporate clean building technologies based on non-fossil-based, renewable energy sources (i.e., wind power, solar power, biomass, hydropower, biofuels) to reduce the use of natural resources, and cut or eliminate emissions and wastes. Today, clean technologies are competitive with their conventional counterparts with low carbon footprints and additional benefits, such as cost and efficiency.

Earth resources   Materials and technologies should be selected from natural or minimally processed earth resources, due to their renewability, low energy use, and low risk of chemical releases during their life cycles. Elements that reduce raw material use should especially be the choice for green buildings, mainly because of their resource conservation. Engineered certified products from renewable resources should be considered as the first choice. For example, a manufacture wood certification, such as Forest Stewardship Council (FSC) is the best way to ensure that the wood material is not produced from natural forests or other habitat around the world. These products contain wood content as low as 17 percent, and the rest of the fiber content is from recycled resources.

Low-embodied energy   Material and technologies that eliminate or reduce energy for excavation, production, manufacturing, construction, and demolition activities (including salvaged and pre- and postconsumer recycled content) should be selected for green buildings. Since local elements do not require long-distance transportation, shipping, and servicing, they are inherently low-embodied products. Special attention should be given to the elements that save energy and water by reducing heating and cooling loads and conserving energy.

Renewable Elements that utilize alternative renewable energy resources, instead of fossil fuels and conventionally generated electricity are highly beneficial from an ecological standpoint. Renewable materials should be selected from products that are made from rapidly growing natural resources. These materials have reduced net emissions of CO_2 across their life cycle and have broad economical benefits.

Biodegradable Biodegradable building materials in green buildings are easiest to recycle and, therefore, can reduce waste, pollution, and energy use. Since these materials come from nature (such as plants and minerals), they are broken down easily by microorganisms, and return to their natural states over time. They are readily disassembled into individual materials. Natural biomaterials such as clay, brick, straw, and additives such as biopolymers and biodegradable resins should be selected as building material choices.

ECOLOGICAL RESOURCES

Sharing the resources among buildings, site selection, soil type, and groundwater conditions must be taken into consideration before the building is designed and constructed. Water and waste-management resources should be ecologically constructed and used.

Resource share Buildings are responsible for one-sixth of the world's fresh water withdrawals, one-quarter of its wood harvest, and two-fifths of its material and energy flows (Roodman and Lenssen 1995). All architectural projects (including buildings, parks, memorials, etc.) should share resources with each other effectively and efficiently.

Green buildings that are connected enjoy an equal share of resources with maximized distribution. For instance, row, attached, and cluster housing share utilities like heat, water, and sewage. Reducing the natural resource consumption should be targeted right from the start, at the design stage. Architects should focus on design issues such as shared outdoor areas, energy, water and sewage networks, the building's frame, and mechanical and electrical services.

Soil/landscape Architects should select the right soil and landscaping for the building type. Soil type and groundwater conditions must be taken into consideration before the building is designed and constructed. Groundwater conditions are important because of their impact on waterproofing as well as structural design. Architects must make sure that water tables underneath the foundation will not be harmed or leaked into/from the sewage system. A high water table may require costly structural and waterproofing techniques and make a site unsuitable.

Construction damage to biotic factors (such as trees and plants) should be minimized. The type and stability of soil should be taken seriously, not only because of possible building damage but also because of potential problems to the soil ecology such as erosion, pollution, sedimentation, and various forms of soil degradation. For example, when too much erosion occurs in a specific area, the water washes away many of the nutrients in the soil. Architects should also avoid converting ecologically productive environments into suburban sprawl or extending out onto land that is reclaimed from the sea (i.e., Hong Kong). Although this action might seem to provide valuable room for development, it results in the loss of rich fishing grounds and ecologically valuable wetlands (Hudson 1979).

Site selection   Site selection is one of the most important resource factors in ecological architecture. Natural qualities, such as water, orientation, vegetation, view, and climate must be considered. Man-made factors like location, utilities, services, other buildings, roads, etc., should be integrated into green building design. Specific site conditions such as steep slopes over 15 percent, severe climatic exposure, earthquake danger areas, flood zones, and unstable soil should be avoided.

Water resources and use   Availability of and access to clean water, along with water conservation, should be the main priorities of green building design. Buildings should avoid excessive use of groundwater for activities other than cleaning and cooking. Architects should install water-saving, ecofriendly devices, recycle gray water for flushing and landscaping, and harvest rainwater. Water-efficient landscaping should be provided. A building site planted with native plants coexists much better with its natural surroundings. Green buildings should use no more than 50 percent of the potable water for irrigation that a typical commercial property of similar size in that area would use. Green buildings should use 30 percent less water than the baseline calculated for traditional buildings (not including irrigation). In order to maximize water efficiency, architects should install water efficient fixtures and appliances.

Waste management   Waste management should be dealt with as an on-going process at different stages throughout the building cycle. A significant amount of waste is generated by the construction of a typical building. Green buildings should be designed to eliminate waste by using modular systems of construction, recycled products, and efficient use of materials. Demolition waste should be recycled by using a percentage of reclaimed materials in other construction projects.

   During the building's life cycle, disposal of all waste should be recycled and treated. All treatment activities must take place within the building with an impermeable surface and sealed drainage system.

ECOLOGICAL ENVIRONMENTS

The prevention of pollution (air, water, and land) is one of the main objectives of green buildings. Global stewardship should be implemented by thinking globally and acting locally. Biodiversification issues should be incorporated into design by preserving existing ecosystems. Planning for responsible land use should address these issues by considering climate, existing ecosystems, and the natural environment.

Pollution control   The materials, technologies, and the type of energy used in the buildings should be selected from nonpollutant elements. Architects should avoid using products that contain compounds that pollute the air—both indoors and out.

Inadequate or faulty pipelines and sewage system installation should be avoided at all costs. Leakage from these systems pollutes soil and groundwater.

Runoff water, including flood control and water supplies, should be managed with a well-designed stormwater system. Green buildings should provide proper drainage systems that collect runoff from impervious surfaces (e.g., roofs and roads) to ensure that water is efficiently conveyed to waterways through pipe networks. Architects should consider using integrated water management (IWM) techniques, including stormwater harvest (to reduce the amount of water that can cause flooding), infiltration (to restore the natural recharging of groundwater), biofiltration or bioretention (e.g., rain gardens) to store and treat runoff to be released at a controlled rate, and wetland treatments (to store and control runoff rates and provide habitat in urban areas).

Global stewardship   Green architects should think globally and act locally. Green buildings should promote and pursue nonrenewable alternative energies, renewable and recyclable materials, and energy- and water-saving technologies. Whereas the material and technological choices should be selected locally, their ecological implications and ramifications of those choices should be considered in the long term and globally.

Biodiverse   Green buildings should be designed to be in contact with nature, integrating biodiverse systems, such as trees, gardens, and green roofs. If artificial ecosystems are introduced, they should be adaptable to their new environment. Green roofs can be designed as part of the building construction, where they provide natural insulation, filter and control stormwater runoff, absorb up to 90 percent of the rainwater, and reduce the urban heat island effect. Landfills and soil erosions can be prevented by displacing soil in different parts of the construction and adding trees, shrubs, and other biotic factors to the site.

Land use   Land use should be planned to save resources and energy use over a large area and to create diverse uses and activities. From an ecological perspective, land use should be responsive to the needs of the building occupants and other members of the community, considering such factors as transportation, infrastructure, and landscape. Trees and other natural features should be protected by developing a preservation plan with no disturbance zone, and by avoiding trenching, grade change, and compacting soil. Undeveloped land should be protected by leaving it in its natural state, not by being excavated or altered. Leaving the site in its natural state allows for stormwater to percolate into the ground, rather than running off into artificial storm drains and costly treatment facilities.

Performance

Performance is defined as the manner of functioning, implying action that follows established patterns or procedures, or fulfills agreed-upon requirements. The term has transitive properties such as execution (carrying out what exists in plan or in intent), discharge (completion of appointed duties and tasks), and fulfillment (complete realization of ends or possibilities). It also has intransitive properties such as accomplishment (successful completion of a process rather than the means of carrying it out).

In architecture, building performance is defined as a measurable outcome of the functional, structural, and environmental qualities of the building. These qualities are measured by determining how well the building supports the needs of its users, including materials and technologies, resources, and environmental behavior of the building. The major concerns for the health, safety, and welfare of the building occupants are addressed by building codes, standards, and regulations (Bullen 2006). As a general framework, green building performance considers all building components during the life cycle by integrating all the subsystems and parts of the building to work together. This is established via three main operational components (see Table 2.3):

1 Performance of the elements (materials and technologies)
2 Resource performance
3 Environment performance

PERFORMANCE OF ELEMENTS

The elements should perform efficiently (in terms of energy, water, cost, and resources), effectively based on their behavioral and physical properties, and productively such as adding value on resources consumed.

Efficiency Green-efficient materials and technologies are high-performance elements that have high source adaptability and are easy to operate, retain, and recover. They use less energy to perform as well or better than traditional elements. They are energy-, water-, cost-, and resource-efficient, reliable elements that produce more with the least possible waste of time and effort, and reduce the overall impact of the built environment on the occupants and the natural environment.

Energy efficiency in green buildings can be accomplished through the reduction of energy consumption by using energy-efficient lighting, heating, and cooling systems; developing strategies for passive (solar) design and natural lighting; and considering alternative and renewable energy sources for energy generation and retention.

Water efficiency can be accomplished by minimizing wastewater with water conserving elements; using self-closing systems and microirrigation to supply water for

TABLE 2.3 COMPONENTS FOR HIGH-PERFORMANCE ARCHITECTURE		
ELEMENTS	**RESOURCES**	**ENVIRONMENT**
Efficiency	Economic	Adaptability
Effectiveness	Ecobehavior	Functionality
Productivity	Design	Environmental quality

landscaping; and using recycled water for toilet flushing or a gray water system that recovers rainwater or other nonpotable water for site irrigation.

Resource efficiency can be accomplished by utilizing locally available elements with recycled, renewable content with minimal waste. These materials and technologies can be reclaimed from disposal and reused by renovating, repairing, restoring, or generally improving the appearance, performance, quality, functionality, and value.

Effectiveness   The effectiveness of elements measures the performance level, which includes all physical, structural, thermal, and behavioral performance factors of the materials and technologies. The results derived from green elements should exceed the actual results of the standard industry elements.

Effective behavioral qualities:
  Strength
  Stability
  Fire/heat resistance
  R-value
  Conduction

Physical properties:
  Structure
  Thermal
  Light
  Electrical

Productivity   The productivity of materials and technologies is measured based on the relationship and ratio between produced result and the resources consumed. A green building should function productively by using energy economically, by protecting the occupants by reacting to environmental and ecological conditions automatically. New smart green materials and technologies benefit the occupants of the building and the environment because of their timely adjustments; e.g., they are constantly monitoring changes and adapting to them. For example, according to Siemen's report, heating and cooling systems, and continuous monitoring and management of lights, coupled with access and control detectors can cut electricity use by 45 percent and reduce energy consumption by 17 percent. When used properly, smart materials can create productive solutions by producing, converting, and storing surplus energy for future use.

RESOURCE PERFORMANCE

Green buildings should contribute to the surrounding area economically, and preserve and protect the ecosystem. They also should incorporate active and passive green strategies, consider market realities, and respect building grammar and human behavior.

Economic   Green buildings should contribute to the economy by providing opportunities for long-term growth for residents, retailers, and commercial tenants, as well as for the entire area. A new green building development can create a community-based economy, where people spend most of their time and can find all the goods and services required to meet their daily needs.

Ecobehavior The ecobehavior of a building can be assessed by measuring its preservation, protection, and enhancement of biodiversity and ecosystems. Ecobehavior can be achieved by improving air and water quality, reducing waste streams, and conserving and restoring natural resources.

Design Green buildings should address the issues of material, technological, environmental, social, and economical aspects of architectural design by balancing active and passive strategies, market realities, building aesthetics, and human behavior. Green building design performance relies on the implementation and coordination of these issues. The separation and/or alienation of any one of these issues will make the green design process incomplete and may produce disappointing results.

ENVIRONMENTAL PERFORMANCE

Green buildings should be adaptable to the climate, and to environmental changes. Architects should ensure the environmental quality for all occupants by providing comfortable, healthy, and habitable buildings.

Adaptability Green buildings should be physically, functionally, and socially adaptable to the environment and perform according to environmental changes. Changes in climate, social patterns, or trends should not end the building's life span, but should rather give birth to different uses for the building.

Functionality The functionality of a green building is determined by the capability of operation of the elements and the resources on the environment. The building should serve well the original function it was designed for. Moreover, multifunctionality adds to the life span and value of a building and the return of equity.

Environmental quality The environmental quality of green buildings should ensure that they are healthy for their occupants, more comfortable, and easier to live in due to lower costs and maintenance requirements. Green buildings should maximize daylighting; have appropriate ventilation and moisture control; and avoid the use of hazardous and toxic materials. Additional consideration must be given to ventilation and filtration to mitigate chemical, biological, and radiological leakage.

Green

Green is an abstract concept, which requires the inclusion of the terms: sustainability, ecology, and performance. Though there is a categorical relationship between the subterms, each category is nevertheless independent and mutually exclusive. For instance, a building can be sustainable but not ecological or green, whereas a green building must be a combination of sustainable, ecological, and performative. The level of greenness is determined based on the level of interaction of these three categories (see Fig. 2.1).

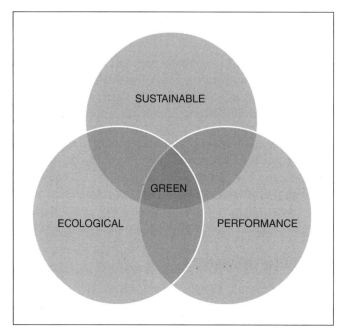

Figure 2.1   Relationship between the green categories.

Measurement of Green

Because of its inherently abstract, subjective nature, measuring an architectural value system is difficult. Measuring the "greenness" of architecture is understandably also challenging. To solve this issue, the three categories that make up the concept of "green"—sustainable, ecological, and performance—have been defined operationally. Linked to each category are the essential components and values that further define and isolate them. Although operational definitions are rarely used in architecture, they are of utmost importance in defining "green" and its relevant categories. These definitions assign meaning to a construct by specifying the activities or "operations" necessary to measure it (Nachmias and Nachmias 1996). According to Bridgman, the meaning of every abstract concept can be measured by assigning quantifiable values to the concept. The concept of "length," therefore, is fixed when the operations by which length is measured are fixed. (That is, the concept of length involves nothing more than the set of operations by which length is determined.) In general, a concept is synonymous with its corresponding set of operations (Bridgman 1980).

Accordingly, as a conceptual framework "green architecture" has three main categories: sustainable, ecological, and performance (see Fig. 2.2). These categories are

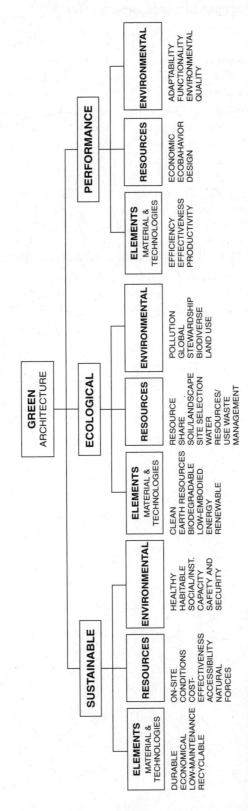

Figure 2.2 Taxonomy of green architecture.

connected but they are also distinct and mutually exclusive. They each have their own subcategories with specific attributes. This taxonomy is not sequential; instead, it works primarily from the bottom up.

For any building to be considered as an example of "green architecture," it should include all of the categories in various degrees. In other words, an architectural artifact can be "green" only if it is simultaneously "sustainable," "ecological," *and* "performative." The level within these categories depends on their own respective taxonomies. Sustainability requires "elements (technology and materials)," "resources," and "environmental" subcategories. The degree of these subcategories is also determined based on their level of inclusion and overlapping attributes. For instance, a building can be "completely" or "barely" sustainable—or nonsustainable—depending on the number of attributes it possess within each subcategory (i.e., elements, resources, and environmental). If a building only uses durable materials, is affordable, and is healthy, then that building is barely sustainable and goes to the bottom of the sustainability scale. If one of these subcategories is missing, then the building cannot be considered sustainable, even though it has sustainable qualities.

BRIEF HISTORY OF
GREEN ARCHITECTURE

Early Beginnings

The history of green architecture is basically the history of mankind. The relationships among man, environment, and ecology were established the day the first humans (hominids) appeared on the scene. Four million years ago, the human evolutionary line had separated from that of the other primates. Hominid development between 4 million BC and 1 million BC falls into two stages: prehuman category (*Australopithecus afarensis*) and human (*Homo habilis* and *Homo erectus*) (Jones, Martin and Pilbeam 1994; Stringer and Andrews 2005).

The first *Homo habilis* lived about two million years ago. They had the ability to make stone tools, which were mainly used to detach meat from animal carcasses. *Homo erectus*, with a brain twice the size of *Homo habilis*, appeared about 1.7 million years ago. They were skillful, innovative, and very adaptable to the environment. Homo erectus invented the hand ax, which was the most versatile and widely used tool for the next million years. The discovery of making and controlling fire gave *Homo erectus* the power to migrate to more northern areas. *Homo erectus* spread beyond its African homeland to the coastal strip of Syria, Lebanon, Arabia, and Iran, and then to China and Southeast Asia (Scarre 1993; Stringer and Andrews 2005; Teeple 2006).

The success of early hominids in spreading beyond the tropical zone is important and remarkable because it took place during a successive cycle of glacial periods and warmer interglacial periods, which began over two million years ago. During the glacial episodes (Ice Ages), which occurred approximately every 100,000 years, ice sheets advanced from the north to cover much of Eurasia and North America. During the interglacial periods, which typically lasted for around 20,000 years, temperatures were similar to, or warmer than, those of today. Within the glacial episodes, there were occasional warmer periods (Scarre 1993). The global ecosystem, including animals and humans, responded to these fluctuating temperatures by spreading northward during milder periods but retreating south again when glaciers advanced. The human race survived these periods by experimenting with new materials, making new tools, and developing new technologies, mainly to fashion clothing and shelters (Basalla 1989; Scarre 1993).

Shelters and Early Technology

Some parts of Europe and Asia offered convenient caves for shelter, but elsewhere humans built windbreaks with stones or made simple shelters of wooden poles, covered with leafy branches or animal hides. Because groups were nomadic, the campsites were temporary, although favored locations were revisited year after year. The all-important stone hand ax was now used for butchering carcasses, working with hides, cutting branches, and preparing plant foods (Basalla 1989).

The period from 100,000 to 35,000 BC was a testing time for the resilience of humans. Despite the bitterly cold conditions of the Ice Age, they not only survived, but also migrated into previously uninhabited regions, including Australia and other remote islands. They developed improved tools and weapons (smaller blades and points set into handles of wood or bone), and composite tools, which were lighter and more efficient than heavier, single-piece counterparts. As the ice sheets began to retreat 15,000 years ago, the world transformed radically. Global temperatures, which had been rising since the coldest point of the last Ice Age (around 20,000 years ago), had almost reached their present level by 8000 BC. As the ice sheets melted, trapped water was released, causing sea levels to rise by around 325 ft. This inundated what had previously been dry lowland areas such as the Bering land bridge, the gulf of Siam, the North Sea, and the Sahul Shelf (between Australia and New Guinea). More water was available to fall as rain, so the extensive deserts that had formed in parts of the tropics began to shrink (Basalla 1989; Scarre 1993; Teeple 2006).

Population Increase

These climatic changes offered new opportunities to the human populations, whose survival strategies had become highly developed under the harsh conditions of the Ice Age. As the climate improved, plants and animals colonized larger areas of the world, including previously uninhabitable parts of Africa, Europe, and North America. The abundant resources in these territories compensated for the lands lost beneath the rising seas. The availability of new food supplies catalyzed a steady growth in human populations, which later caused the number of the hunter tribes to increase to unsustainable levels. The gathering and hunting way of the nomads became almost impossible because of their population size and their increased need for food. Therefore, around 9000 BC, hunter-gathering tribes started to form settlements and farming communities. Whole communities could be supported by cultivated crops, and storage of grain throughout the year made it easier for people to live in fixed settlements. The nomadic life of hunter-gatherer groups gradually became more settled as dependence on different cultivated plants increased.

Ecological/Environmental Changes

The transformation from nomadic to agrarian society was a paradigm shift, where human intervention with the environment was followed by ecological changes. The emergence of new settlements demanded resources, and forests were the first to suffer, as the requirements for building, heating, and cooking rose steadily. Local deforestation and consequent soil erosion started to decay the environment and caused major ecological problems around settled areas (Ponting 2007). The first signs of ecological damage emerged in Central Jordan in 6000 BC. Villages were abandoned as soil erosion caused by deforestation resulted in a badly damaged landscape, declining crop yields, and eventual inability to grow enough food (Ponting 2007; Montgomery 2008). Between 1300 and 900 BC, there was an agricultural collapse in the central area of Mesopotamia following salinization as a result of too much irrigation. In Rome, 312 BC, heavy water pollution was reported in the Tiber River. The Roman senate pronounced the river too polluted to use as drinking water and constructed the first aqueduct (Ponting 2007).

Ecological changes due to deforestation were seen throughout the world. In Japan, the rise of the first settled societies and then the building of towns resulted in steady deforestation, which in turn caused disastrous flooding of rivers around 1600 BC. This chain reaction of environmental destruction led to the first recorded governmental legislation to control the forests (Ponting 2007). In 146 BC, North Africa, a growing Roman demand for grain pushed cultivation further into the hills and onto vulnerable soils that were easily eroded when deforested. In Asia Minor, 1 BC, the interior of the old Roman province Phrygia was completely deforested and, as a result, access to the remaining forests of Asia Minor and Syria was restricted by Rome. In the Americas in AD 600, soil erosion caused by deforestation caused the entire Mayan agricultural base to collapse. As a result, within a few decades all Mayan cities were abandoned.

These changes accelerated after AD 1100 with global population surges and the rise of Europe. The population increase led to massive movements of people and increasing demand for energy. Millions of acres of additional forests were destroyed to increase the area of habitable land and to satisfy the increasing demand for timber, which was the main raw material of the time (Gimpel 1976; Gimpel 1979). This soon necessitated a search for new sources of energy. Unfortunately, the use of coal as a substitute fuel for wood and the subsequent industrialization created an ecological disaster. Significant ecological changes, destruction of arable lands, and massive environmental pollution were seen at every level and in every place.

Descriptions of these destructions were reported in England in AD 1230, 1236, 1257, and 1835, in Japan in 1610 and 1870, and in France in 1350, 1560, and 1852. The heavy pollution made the air unbreathable, rivers unusable, and groundwater undrinkable. Numerous deadly diseases were spawned, triggering governmental legislations. In London in 1287, a commission was set up to investigate complaints about smoke levels. In 1307, the burning of coal in London was banned. In 1366, Paris butchers were forced to move out of the city and to dump their garbage in the rural streams. In 1582, heavy water pollution caused Dutch authorities to order linen bleachers not to dump their wastes into the canals, but to use separate disposal channels. In 1860, Norway made an official complaint about the acid rain coming from Britain (Gimpel 1976; Gimpel 1979; Ponting 2007; Montgomery 2008).

The Rise of Environmental Awareness

Environmental awareness emerged only after ecological destruction caused insupportable environmental changes. Most of these early reactions appeared as anecdotes and local legislative environmental policies (Opie 1998). The first large-scale, modern environmental legislations came with the British Alkali Acts, passed in 1863 to regulate the toxic air pollution caused by the production of soda ash. Thus environmentalism began as a reaction to the growth of cities, industrialization, and worsening air and water pollution.

In the United States, environmental movement can be traced as far back as 1739, when Benjamin Franklin and other Philadelphians, citing "public rights," petitioned the Pennsylvania Assembly for legislation to stop the dumping of waste and to remove tanneries from Philadelphia's commercial district (Isaacson 2004). However, during the nineteenth century, a wave of appreciation for nature swept across America. This appreciation began as recognition of the relationship between nature and man, which were reflected in the literature and artistic works of the time, such as popular landscape paintings of the American West by Albert Bierstadt and Edwin Church (Fox 1986; Nash 2001; Warren 2003; Merchant 2005).

Many writers also romanticized or focused upon nature with a concern for protecting the natural resources of the West, with individuals such as Henry David Thoreau, George Perkins Marsh, and John Muir making key philosophical contributions. Thoreau was interested in the relationship of people with nature and concerned about the wildlife in Massachusetts; he wrote *Walden; or, Life in the Woods* (Thoreau 1854) as he spent two years living in a cabin near Walden Pond (see Fig. 3.1). In 1864, George Perkins Marsh, who was concerned with the need for resource conservation,

WALDEN;

OR,

LIFE IN THE WOODS.

By HENRY D. THOREAU,
AUTHOR OF "A WEEK ON THE CONCORD AND MERRIMACK RIVERS."

I do not propose to write an ode to dejection, but to brag as lustily as chanticleer in the
morning, standing on his roost, if only to wake my neighbors up. — Page 92.

BOSTON:
TICKNOR AND FIELDS.
M DCCC LIV.

Figure 3.1   Thoreau's book, *Walden or Life in the Woods* is considered as the first published environmentalist book in the United States in 1854.

published his very influential book *Man and Nature*. In it Marsh argued that deforestation could lead to desertification, the gradual transformation of habitable land into desert (Marsh 2006).

Conservation Movement

These efforts set the tone for the early Conservation Movement in the United States. An expedition into northwest Wyoming in 1871 led by Hayden and Jackson provided the imagery needed to substantiate stories about the almost mythical Yellowstone region. His report helped to persuade the U.S. Congress to open the area for public settlement; on March 1, 1872, President Ulysses S. Grant signed the Act of Dedication into law, creating Yellowstone National Park.

Conservation efforts were further promoted and publicized by John Muir, who relentlessly petitioned Congress to pass a bill turning the region into a national park. Finally, the U.S. Congress passed a legislation to reserve the area soon to be known as Yosemite National Park on October 1, 1890 (Schaffer 2000). A decade later, Presidents Harrison, Cleveland, and McKinley had transferred approximately 50 million acres into the forest reserve system. During the Conservation Movement, President Roosevelt's contribution cannot be neglected. In 1902, he signed the National Reclamation Act, which funded irrigation projects for large tracts of arid land in the West. In 1905, President Roosevelt again helped to create the United States Forest Service and by the end of his presidency, he had successfully increased the number of national parks as well as added acreage to existing forest reserves.

The Birth of Modern Environmentalism and Specialized Movements

The conservationist principles, based on the belief that nature has an inherent right, were to become the foundation of modern environmentalism in the twentieth century. Environmental awareness continued to increase in popularity and recognition, but only after World War II did a wider awareness began to emerge. Although the intent was the same, the concerns and the issues of modern environmentalism were more global and complex than mere nature conservationism. Scientific discoveries, advancements in technologies and materials, increasing ecological stress and global pollution, growing population, limited resources, and economic disparities among the nations became the new agenda of the environmental movement. Environmentalism was still the general framework, but new specialized movements began to emerge.

ECOLOGICAL MOVEMENT AS ENVIRONMENTALISM

The ecological movement was the first branch to appear. The movement's growth was stimulated by a widespread acknowledgement of problems with the ecosystem, natural resources, and pollution. One of the first publications of the ecological movement was Aldo Leopold's book, *Sand County Almanac* (Leopold 1949), which emphasized the morality of respect for the environment and the ecological importance of conservation. Although it was written mainly from a conservationist viewpoint, the major contribution of this book was presenting ecology as a science.

A major cornerstone for the ecological movement was the publication of Rachel Carson's book *Silent Spring*, which drew significant attention to the impact of chemicals on the natural environment (Carson 1962). *Silent Spring*, released in 1962, offered the first look at widespread ecological degradation and triggered the environmental awareness that exists today. This book focused on the harmful effects of insecticides, and other chemical products, as well as the use of sprays in agriculture. Supported with solid evidence, this book is considered to be the starting point of the modern environmental movement in the United States, mainly because of the following environmental legislations of the 1970s.

In 1964, the Wilderness Act was written by the Wilderness Society and signed into law by President Johnson. It created the base definition of wilderness in the United States and protected over nine million acres of federal land. This federal law represented a significant achievement for the Ecological Movement. For almost four centuries, the United States had promoted a philosophy of conquering the wilderness, but this act committed society to preserving the wilderness. The underlying concept of self-control for nature preservation was imperative for ecological protection and maturity.

In the United States, two important incidents were turning points for the Ecological Movement. They were: (1) the Sierra Club's 1966 campaign against the Bridge Canyon Dam, which resulted in the cancellation of the dam's construction plans in 1968; and (2) the Santa Barbara oil spill in 1969. These incidents outraged the public, further increased ecological awareness, and subsequently helped to turn it into a social movement. In response, the government passed a series of environmental legislations, namely:

- 1969: National Environmental Policy Act
- April 22, 1970: Earth Day was founded by U.S. Senator Gaylord Nelson as an environmental teach-in and to inspire awareness of and appreciation for the earth's environment.
- 1970: Creation of the Environmental Protection Agency (EPA)
- 1970: Clean Air Act (revised and reintroduced in 2006)
- 1972: Clean Water Act (Federal Water Pollution Control Amendment)

These ecological legislations were later followed by the enactment of a whole series of laws regulating waste: the Resource Conservation and Recovery Act in 1976, the Toxic Substances Control Act in 1976, the Federal Insecticide, Fungicide, and Rodenticide Act (FIFRA) in 1972 (as a revision of the Federal Insecticide Act of 1910), and for the cleanup of polluted sites, the Comprehensive Environmental Response, Compensation, and Liability Act (CERCLA) in 1980.

The Oil Embargo of 1973 and subsequent energy crisis led to greater interest in renewable energy and spurred research in alternative energy sources, including solar, wind, and nuclear power. As a result of public pressure, President Carter signed the National Energy Act (NEA) and the Public Utilities Regulatory Policy Act (PURPA). In 1977, President Carter convinced Congress to create the United States Department of Energy (DoE) with the goal of conserving energy. That same year, Congress enacted federal solar tax credits as a subsidy to lower the cost of solar energy systems, and Carter had solar hot water panels installed on the roof of the White House.

The architecture field followed the trend closely. A number of architects adopted new energy strategies, and hundreds of solar homes appeared across America. Ecologically sound practices like improved insulation, energy efficiency, and solar energy were incorporated into the architecture of the day. In 1976, Naar's book, *Design for a Limited Planet*, (Naar 1976) reviewed alternative energy houses. Due to the changes in consumer preferences and demand, the renovation of old buildings and reusing existing structures emerged as a new trend in architecture. Empty warehouses, train stations, abandoned factories, and office buildings were reutilized as mix-use developments.

SUSTAINABILITY MOVEMENT AS ENVIRONMENTALISM

During the 1960s, increasing concerns about economic growth and its environmental effects brought environmental "sustainability" issues to the surface. In 1968, Paul Ehrlich raised these concerns in his book, *Population Bomb* (Ehrlich 1968), describing the economic and ecological threats of a rapidly growing human population. Ehrlich argued that this growth rate was unsustainable, and an unprecedented collective effort was needed to return human use of natural resources to within sustainable limits (Ehrlich 1968; Dresner 2002). This was the first time the concept of "sustainability" emerged as part of the modern environmental movement. Triggered by various conferences, including the Ecological Aspects of International Development Conference in Washington, D.C. (1968), the UNESCO Biosphere Conference in Paris (1968), and the United Nations Conference on the Human Environment (1972), sustainability became increasingly associated with the integration of economic, social, and environmental issues (Barrow 1995). Sustainability was first formally introduced as a concept in "Our Common Future," the proceedings of the Brundtland Commission. The Brundtland Report, which defined the term as "meeting the needs of the present without compromising the ability of future generations to meet their own needs" (WCED 1987), was primarily concerned with securing global equity, redistributing resources toward poorer nations while encouraging their economic growth. The report also recognized that achieving this goal and sustainable growth would require technological and social change.

The Brundtland Report highlighted three fundamental components for sustainable development: (1) environmental protection, (2) economic growth, and (3) social equity. The Commission outlined a series of "strategic imperatives," or "critical objectives," inherent in their concept of sustainable development (WCED 1987). These included:

- Conserving and enhancing the resource base
- Meeting essential needs (food, water, energy, sanitation, and jobs)
- Merging environment and economics in decision making
- Reviving growth
- Changing the quality of growth
- Reorienting technology and managing risk
- Ensuring a sustainable level of population

The idea of sustainable development gained momentum from the 1992 UN Conference on Environment and Development (UNCED), held in Rio de Janeiro. Shortly after this event, the Earth Council was developed as a nongovernment organization (NGO) to coordinate the efforts of all nations to achieve sustainable development through the creation of national councils. The Rio Conference produced three major documents: (1) Convention on Biological Diversity; (2) Framework Convention on Climate Change; and (3) Agenda 21. Agenda 21 was a policy declaration, or Action Plan; the other two documents were legally binding international treaties.

The importance of Agenda 21 is that it sets forth specific policy recommendations for sustainable development. After his election in 1992, President Clinton complied with the recommendation of Agenda 21 by issuing Executive Order 12852, which created the President's Council on Sustainable Development (PCSD). Agenda 21 identifies 27 guiding principles, known as "The Rio Declaration," in order to integrate three principles into all public policy: (1) economic growth; (2) environmental protection; and (3) social equity.

Agenda 21 Principles (UN 1992):

Principle 1
Human beings are at the core of concerns for sustainable development. They are entitled to a healthy and productive life in harmony with nature.

Principle 2
States have, in accordance with the Charter of the United Nations and the principles of international law, the sovereign right to exploit their own resources pursuant to their own environmental and developmental policies, and the responsibility to ensure that activities within their jurisdiction or control do not cause damage to the environment of other States or of areas beyond the limits of national jurisdiction.

Principle 3
The right to development must be fulfilled so as to equitably meet developmental and environmental needs of present and future generations.

Principle 4
In order to achieve sustainable development, environmental protection shall constitute an integral part of the development process and cannot be considered in isolation from it.

Principle 5
All States and all people shall cooperate in the essential task of eradicating poverty as an indispensable requirement for sustainable development, in order to decrease the disparities in standards of living and better meet the needs of the majority of the people of the world.

Principle 6
The special situation and needs of developing countries, particularly the least developed and those most environmentally vulnerable, shall be given special priority.

International actions in the field of environment and development should also address the interests and needs of all countries.

Principle 7
States shall cooperate in a spirit of global partnership to conserve, protect and restore the health and integrity of the Earth's ecosystem. In view of the different contributions to global environmental degradation, States have common but differentiated responsibilities. The developed countries acknowledge the responsibility that they bear in the international pursuit of sustainable development in view of the pressures their societies place on the global environment and of the technologies and financial resources they command.

Principle 8
To achieve sustainable development and a higher quality of life for all people, States should reduce and eliminate unsustainable patterns of production and consumption and promote appropriate demographic policies.

Principle 9
States should cooperate to strengthen endogenous capacity-building for sustainable development by improving scientific understanding through exchanges of scientific and technological knowledge, and by enhancing the development, adaptation, diffusion and transfer of technologies, including new and innovative technologies.

Principle 10
Environmental issues are best handled with the participation of all concerned citizens, at the relevant level. At the national level, each individual shall have appropriate access to information concerning the environment that is held by public authorities, including information on hazardous materials and activities in their communities, and the opportunity to participate in decision-making processes. States shall facilitate and encourage public awareness and participation by making information widely available. Effective access to judicial and administrative proceedings, including redress and remedy, shall be provided.

Principle 11
States shall enact effective environmental legislation. Environmental standards, management objectives and priorities should reflect the environmental and developmental context to which they apply. Standards applied by some countries may be inappropriate and of unwarranted economic and social cost to other countries, in particular developing countries.

Principle 12
States should cooperate to promote a supportive and open international economic system that would lead to economic growth and sustainable development in all countries, to better address the problems of environmental degradation. Trade policy measures for environmental purposes should not constitute a means of arbitrary or unjustifiable discrimination or a disguised restriction on international trade. Unilateral

actions to deal with environmental challenges outside the jurisdiction of the importing country should be avoided. Environmental measures addressing transboundary or global environmental problems should, as far as possible, be based on an international consensus.

Principle 13
States shall develop national law regarding liability and compensation for the victims of pollution and other environmental damage. States shall also cooperate in an expeditious and more determined manner to develop further international law regarding liability and compensation for adverse effects of environmental damage caused by activities within their jurisdiction or control to areas beyond their jurisdiction.

Principle 14
States should effectively cooperate to discourage or prevent the relocation and transfer to other States of any activities and substances that cause severe environmental degradation or are found to be harmful to human health.

Principle 15
In order to protect the environment, the precautionary approach shall be widely applied by States according to their capabilities. Where there are threats of serious or irreversible damage, lack of full scientific certainty shall not be used as a reason for postponing cost-effective measures to prevent environmental degradation.

Principle 16
National authorities should promote the internalization of environmental costs and the use of economic instruments, taking into account the approach that the polluter should, in principle, bear the cost of pollution, with due regard to the public interest and without distorting international trade and investment.

Principle 17
Environmental impact assessment, as a national instrument, shall be undertaken for proposed activities that are likely to have a significant adverse impact on the environment and are subject to a decision of a competent national authority.

Principle 18
States shall immediately notify other States of any natural disasters or other emergencies that are likely to produce sudden harmful effects on the environment of those States. Every effort shall be made by the international community to help States so afflicted.

Principle 19
States shall provide prior and timely notification and relevant information to potentially affected States on activities that may have a significant adverse environmental effect and shall consult with those States at an early stage and in good faith.

Principle 20
Women have a vital role in environmental management and development. Their full participation is therefore essential to achieve sustainable development.

Principle 21
The creativity, ideals and courage of the youth of the world should be mobilized to forge a global partnership in order to achieve sustainable development and ensure a better future for all.

Principle 22
Indigenous people and their communities and other local communities have a vital role in environmental management and development because of their knowledge and traditional practices. States should recognize and duly support their identity, culture and interests and enable their effective participation in the achievement of sustainable development.

Principle 23
The environment and natural resources of people under oppression, domination and occupation shall be protected.

Principle 24
Warfare is inherently destructive of sustainable development. States shall therefore respect international law providing protection for the environment in times of armed conflict and cooperate in its further development, as necessary.

Principle 25
Peace, development and environmental protection are interdependent and indivisible.

Principle 26
States shall resolve all their environmental disputes peacefully and by appropriate means in accordance with the Charter of the United Nations.

Principle 27
States and people shall cooperate in good faith and in a spirit of partnership in the fulfillment of the principles embodied in this Declaration and in the further development of international law in the field of sustainable development.

GREEN MOVEMENT AS ENVIRONMENTALISM

The "green movement," with deep philosophical roots and strong political connotations, has emerged as a combination of all the previous environmentalist movements. In the early 1970s, pressing issues like economic stability, unemployment, energy, pollution, and social change created a grassroots movement with a strong ideology of green politics. Although the U.S. environmental movement, arising from the late nineteenth century conservationism, had deeper roots than in any other country, the recognition of the need for a fundamental structural change came from abroad (Carter 2001; Talshir 2002).

Various green political parties and organizations were founded in Europe and New Zealand almost a decade earlier than in the United States. The first green political organization, Greenpeace, was formed in Canada in 1971, followed by the first green political party, the Values Party, in New Zealand in 1972, and the first European Green political party in the United Kingdom in 1973. Because of its ecological roots, the UK's green party was named the Ecology Party, which published the first edition of the "Manifesto for a Sustainable Society," arguing that societies must be completely restructured in order to improve the health of the planet (Talshir 2002).

In 1979, the German Green Party, die Grunen, published "The Four Pillars of the Green Party" statement and formed the basis of the worldwide Green Party movement. The four pillar manifesto consisted of (1) Ecological Wisdom; (2) Social Justice; (3) Grassroots Democracy; and (4) Nonviolence (Talshir 2002). In 1984, the U.S. Green Committee expanded these four pillars into the Ten Key Values, namely:

- Grassroots Democracy
- Social Justice and Equal Opportunity
- Ecological Wisdom
- Nonviolence
- Decentralization
- Community-Based Economics and Economic Justice
- Feminism and Gender Equity
- Respect for Diversity
- Personal and Global Responsibility
- Future Focus and Sustainability

The 1980s saw various attempts to address these issues, including economic sustainability, population control, pollution, and ecological destruction. The United Nation's reports in 1987 and 1992, numerous ecological disasters such as oil spills in the late 1980s and early 1990s, and the warnings by scientists about global warming, have received overwhelming attention by the public. As a result, a hybrid movement started to emerge: sustainability, ecology, and green movements began to overlap, and these three terms are still used interchangeably today.

Although the architectural discipline has been directly correlated with these developments, a formal green building field did not develop until the 1970s. The American Institute of Architects (AIA) Energy Committee was founded in 1973 to study energy issues in buildings and to develop regional guidelines for passive solar design and building energy performance standards. As an extension of this committee, AIA formed the Committee on the Environment (COTE) in 1990, "to advance, disseminate, and advocate—to the profession, the building industry, the academy, and the public—design practices that integrate built and natural systems and enhance both the design quality and environmental performance of the built environment" (AIA—COTE Mission, 2009). Accordingly, COTE developed "Ten Measures of Sustainable Design" principles:
Ten Measures of Sustainable Design (AIA—COTE Mission, 2009) (www.aia.org):

- Design and Innovation
- Sustainable design is an inherent aspect of design excellence. Projects should express sustainable design concepts and intentions and take advantage of innovative programming opportunities.

- Regional/Community Design
- Sustainable design values the unique cultural and natural character of a given region.
- Land Use and Site Ecology
- Sustainable design protects and benefits ecosystems, watersheds, and wildlife habitat in the presence of human development.
- Bioclimatic Design
- Sustainable design conserves resources and maximizes comfort through design adaptations to site-specific and regional climate conditions.
- Light and Air
- Sustainable design creates comfortable interior environments that provide daylight, views, and fresh air.
- Water Cycle
- Sustainable design conserves water and protects and improves water quality.
- Energy Flows and Energy Future
- Sustainable design conserves energy and resources and reduces the carbon footprint while improving building performance and comfort. Sustainable design anticipates future energy sources and needs.
- Materials and Construction
- Sustainable design includes the informed selection of materials and products to reduce product-cycle environmental impacts, improve performance, and optimize occupant health and comfort.
- Long Life, Loose Fit
- Sustainable design seeks to enhance and increase ecological, social, and economic values over time.
- Collective Wisdom and Feedback Loops
- Sustainable design strategies and best practices evolve over time through documented performance and shared knowledge of lessons learned.

GREEN BUILDING RATING SYSTEMS

The development of building rating systems was the result of growing concerns in the building industry and management, in topics such as sustainability, building performance, environmental impact, energy, cost efficiency, and maintenance. The rating systems were a partial response to these issues, proposing quantifiable tools to evaluate and measure the level of a building's environmental performance. Several countries created their own standards of building performance, evaluation, and rating systems, addressing a wide range of environmental issues (i.e., energy, design, construction, site, technologies, and materials). In 1981, the R-2000 building evaluation program was created as a partnership between the Canadian Home Builders' Association and Natural Resources Canada, promoting sustainable technologies in residential buildings. In 1990, the Building Research Establishment's Environmental Assessment Method (BREEAM) was one of the first acknowledged rating systems to evaluate the sustainability of new office buildings in the United Kingdom. In 1992, the Environmental Resource Guide was published by AIA, and in the same year the EPA and the U.S. Department of Energy launched the Energy Star Program. In 1993, the U.S. Green Building Council (USGBC) was founded, and in 1998 they launched their Leadership in Energy and Environmental Design (LEED) pilot program. Currently, there are more than two dozen building rating systems worldwide, and they can be broadly categorized into one of three systems (Trusty 2000; Chew and Das 2004):

- Decision-making support systems (Demkin 1996), (Building for Environmental and Economic Sustainability: BEES; Australian Building Greenhouse Rating: ABGR)
- Performance-based evaluation systems (NABERS 2005)
- Whole building evaluation systems (Liu, Prasad, Li, Fu and Liu 2006)

Green building rating systems often use different evaluation criteria, methods, and procedures, ranging from scoring to categorization. As such, each system has its own advantages and disadvantages. There have been several studies in recent years that evaluated and/or compared many of these systems individually or to each other. These studies have focused on: green building rating systems (Chew and Das 2004; Gowri 2004; Fowler and Rauch 2006; Hubbard 2009); marketability, trends, and applications (Crawley and Aho 1999); building categories and types (Horvat and Fazio 2005); and international market applications (Anneling 1998; Nie, Qin, Jiang and Song 2002; Moro 2004; Lee and Burnett 2006; Chang, Chiang and Chou 2007).

Regardless of their differences, however, most of these systems utilize similar criteria to evaluate and rate green buildings, namely:

- Design
- Construction
- Site
- Indoor air quality
- Environmental impact
- Technology
- Materials
- Energy and water consumption

The following Table 3.1 gives the primary green building rating systems.

TABLE 3.1   PRIMARY GREEN BUILDING RATING SYSTEMS

ABBR	NAME	ORIGIN	GOVERNANCE	EST. YEAR	SCOPE	WEB SITE
R-2000	R-2000	Canada	Canadian Home Builders' Association (CHBA) and the Office of Energy Efficiency (OEE) of Natural Resources Canada (NRCan)	1981	The R-2000 Standard is a series of technical requirements for energy efficiency, environmental responsibility, and new home performance that supplement the existing building codes. Every R-2000 home is built and certified to this standard.	http://r2000.chba.ca
BREEAM	Building Research Establishment's Environmental Assessment Method	UK	BRE (Building Research Establishment)	1990	BREEAM checks wide-ranging environmental and sustainability issues by providing building performance evaluations in nine distinct categories: (1) management; (2) health; (3) well-being; (4) energy; (5) transport; (6) water; (7) material and waste; (8) land use and ecology; and (9) pollution. BREEAM uses a scoring system based on scale of pass, good, very good, excellent, and outstanding. It also uses a number-based rating of 1–5 stars in some regions (i.e., Gulf).	http://www.breeam.org http://www.bre.co.uk
SBTOOL (formerly known as GBTool)	Sustainable Building Tool	Canada	Green Building Challenge (GBC)	1996	SBTool is a customizable building rating system, which evaluates environmental and sustainability performance. The system is designed as a generic toolbox, which can be customized according to local and regional	http://www.greenbuilding.ca

Abbreviation	Name	Country	Organization	Year	Description	URL
					building performance requirements and needs. SBTool uses a scoring system based on a scale of: −1 (deficient), 0 (minimum pass), +3 (good practice), and +5 (best practice).	
LEED	Leadership in Energy and Environmental Design	USA	US Green Building Council (USGBC)	1998	LEED provides a rating framework for developing and evaluating high-performance green buildings. The system primarily measures six categories: (1) sustainable site development, (2) water efficiency, (3) energy and atmosphere, (4) materials and resources, (5) indoor environmental quality, and (6) innovation and design process. LEED uses a 69-point scale system with four ratings: platinum (52–69 pts), gold (39–51), silver (33–38), and certified (26–32).	http://www.usgbc.org/leed
CHPS	Collaborative for High Performance Schools	USA	The Collaborative for High Performance Schools	1999	CHPS facilitates the design, construction, and operation of high-performance school buildings and environments. CHPS's main criterion is to create sustainable school environments, which are not only energy- and resource-efficient but also healthy, habitable, and comfortable.	http://www.chps.net/
GREEN GLOBES	Green Globe Go Green (for existing buildings) is owned and operated by	USA Canada	In USA: Green Building Initiative (GBI) (2005). In Canada: ECD Energy and Environment Canada	2000	Green Globes (GG) is an environmental building design and management tool. It provides an online assessment protocol, a rating	http://www.green-globes.com

TABLE 3.1 PRIMARY GREEN BUILDING RATING SYSTEMS (CONTINUED)

ABBR	NAME	ORIGIN	GOVERNANCE	EST. YEAR	SCOPE	WEB SITE
			and BOMA Canada under the brand name "Go Green" (Go Green Plus) (2004). The Green Globes system has also been used by the Continental Association for Building Automation (CABA) to power a building intelligence tool, called Building Intelligence Quotient (BIQ).		system, and offers guidance for green building design, operation, and management. GG is interactive, flexible, and generates assessment and guidance reports. The system is reportedly better suited for smaller buildings, and serves as an evaluation tool during the design process. It addresses energy, water, waste, resource use, site, hazardous materials, management, health and safety, and indoor environment. Green Globes has two assessment levels: self-assessment and third-party verified assessment. Within its framework, there are modules for each stage of the design process (predesign, design, construction, commissioning).	
BEAM	Building Environmental Assessment Method	Hong Kong	Business Environment Council (BEC), and HKBeam Society	2002	BEAM evaluates and measures the environmental performance of buildings in Hong Kong. The evaluation is based on five building performance criteria: (1) hygiene, health, comfort, and amenity; (2) land use, site impacts, and transportation; (3) use of materials, recycling,	http://www.hk-beam.org.hk

Note: The NAME column also contains continuation text: "BOMA Canada. All other Green Globes products in Canada are owned and operated by ECD Energy and Environment, Canada. The Green Building Initiative owns the license to promote and further develop Green Globes in the United States. GBI is an accredited standards developer under the American National Standards Institute (ANSI), and has begun the process to establish Green Globes as an official ANSI standard."

GGHC	Green Guide for Health Care	USA	American Society for Healthcare Engineering (ASHE)	2003	and waste management; (4) water quality, conservation, and recycling; and (5) energy efficiency, conservation, and management. BEAM uses an overall assessment rating system based on a gained credit percentage scale. Accordingly, BEAM awards four rating classifications: platinum (excellent, 75%), gold (very good, 65%), silver (good, 55%), and bronze (above average, 40%). GGHC is the healthcare sector's first quantifiable, sustainable design evaluation tool. It integrates environmental and health principles and practices into the planning, design, construction, operations, and maintenance of healthcare facilities. In addition to specialized guidelines and evaluation procedures, GGHC uses the LEED system as their existing building-rating mechanism.	http://www.gghc.org
GREEN STAR	Green Star Building Evaluation System	Australia	Green Building Council of Australia Green Star New Zealand Green Star South Africa		Green Star is a comprehensive, national, voluntary rating system that evaluates a building's environmental design and performance. Green Star is modeled after BREEAM; it uses a customizable rating tool kit that can be modified for different building types and functions. Green Star ratings are based on a percentage score	http://www.gbca.org.au/green-star

TABLE 3.1 PRIMARY GREEN BUILDING RATING SYSTEMS (CONTINUED)

ABBR	NAME	ORIGIN	GOVERNANCE	EST. YEAR	SCOPE	WEB SITE
					across nine performance categories: (1) management; (2) indoor environment; (3) energy; (4) transportation; (5) water; (6) materials; (7) land use and ecology: (8) emissions; and (9) innovation.	
CASBEE	Comprehensive Assessment System for Building Environmental Efficiency	Japan	Japan GreenBuild Council (JaGBC)/Japan Sustainable Building Consortium (JSBC)	2004	CASBEE measures the sustainability and environmental efficiency of high-performance buildings. Green building issues and problems that are unique to Japan and Asia are especially taken into consideration. CASBEE has four grading categories (predesign, new construction, existing buildings, renovations), which are evaluated based on four criteria: (1) energy: (2) site; (3) indoor environmental quality; and (4) resource, materials, and water conservation. An overall evaluation rating is determined based on numerous calculations, and the results are presented in a letter scale of S (excellent), A, B+, B−, and C (poor).	http://www.ibec.or.jp/CASBEE/english
HQE	Haute Qualité Environnementale (High Quality	France	Association pour la Haute Qualité Environnementale (ASSOHQE)	2004	HQE evaluates the environmental impact of buildings, focusing on the following criteria: (1) design;	http://www.assohqe.org

	Environmental Standard)				(2) construction; (3) energy; and (4) water, waste, and maintenance.	
NGBS	National Green Building Standard	USA	NAHB (National Association of Home Builders)	2005	NGBS provides guidelines for the mainstream homebuilder to incorporate environmental concerns into a new home. Divided into two parts, the system covers seven evaluation criteria: (1) lot design; (2) resource efficiency; (3) energy efficiency; (4) water efficiency, (5) indoor environmental quality; (6) homeowner education; and (7) global impact.	http://www.nahbgreen.org
NABERS	National Australian Built Environment Rating System	Australia	New South Wales Government (NSW), Department of Environment and Climate Change, Australia	2005	NABERS measures existing buildings' environmental performance during their life cycles. There are separate ratings for: office buildings, office occupants, hotels, and homes. Final ratings are based on measured operational impacts of four evaluation criteria: (1) energy; (2) water; (3) waste management; and (4) indoor environment. NABERS uses the Green Star rating system for final evaluations.	http://www.nabers.com.au
GRIHA	India Green Rating for Integrated Habitat Assessment	India	The Energy and Resources Institute (TERI)/Green Rating for Integrated Habitat Assessment, India	2005	GRIHA measures the environmental performance of buildings, focusing on India's varied climate and building practices. The rating is based on quantitative and qualitative assessment techniques, and is applicable to new	http://www.teriin.org

TABLE 3.1 PRIMARY GREEN BUILDING RATING SYSTEMS (CONTINUED)

ABBR	NAME	ORIGIN	GOVERNANCE	EST. YEAR	SCOPE	WEB SITE
					and existing buildings (commercial, institutional, and residential). The evaluation criteria are: (1) site planning; (2) building envelope design; (3) building system design; (4) HVAC; (5) lighting and electrical; (6) integration of renewable energy sources to generate energy onsite; (7) water and waste management; (8) selection of ecologically sustainable materials; and (9) indoor environmental quality.	
BCA Green Mark	Singapore Building and Construction Authority (BCA) Green Mark	Singapore	Singapore Building and Construction Authority/National Environment Agency	2005	BCA Green Mark is a green building rating system that evaluates a building for its environmental impact and performance. It provides a comprehensive framework for assessing building performance and environmental friendliness. Buildings are awarded the BCA Green Mark based on five key criteria: (1) energy efficiency; (2) water efficiency; (3) site/project development, and management (building banagement and operation for existing buildings); (4) indoor environmental quality and environmental protection; and (5) innovation.	http://www.bca.gov.sg/ GreenMark/green_ mark_buildings.html

Abbreviation	Name	Country	Agency	Year	Description	URL
Three STAR	Green Building Evaluation Standard	China	Ministry of Construction/Ministry of Housing and Urban Rural Development	2006	As China's first building rating system, Three STAR is designed to create local building standards. It is a credit-based system with two standards—residential and commercial. The system evaluates the buildings in six categories: (1) land savings and outdoor environment; (2) energy saving; (3) water savings; (4) material savings; (5) indoor environmental quality; and (6) operations and management.	http://www.cin.gov.cn/
GBAS	Green Building Assessment System	China	Ministry of Science and Technology (MoST)	2006	GBAS is developed from China's Green Olympic Building Assessment System (GOBAS, 2003), and measures basic environmental performance of buildings such as: electricity, water, and energy consumption.	
GSBC	German Sustainable Building Certification	Germany	German Sustainable Building Council (DGNB)	2009	GSBC is a comprehensive rating system, which covers all relevant topics of sustainable building and construction. It was created as a tool for the planning and evaluation of buildings, using six categories with 49 criteria: (1) ecological quality; (2) economical quality; (3) socio-cultural and functional quality; (4) technical quality; (5) quality of the processes; and (6) site. GSBC uses a number-based system, in which each category has an equal percentage weight. Three evaluation degrees are offered: gold (89%), silver (69%), and bronze (50%).	http://www.dgnb.de/en

GREEN TECHNOLOGIES:

ENERGY GENERATION

4

Technology is defined as a practical application of (technical) knowledge, which requires the use of tools and crafts to control and adapt our environment. The term originates from the Greek "techne" (craft) and "logia" (saying); however, its current use is inherently broad and elusive. It can refer to material objects (such as machines) or to broader knowledge areas (such as systems and methods). Today, most dictionaries define technology as an "applied science," assuming the two terms have a causal relationship. Although these terms often seem to overlap (especially since the Industrial Revolution), there is actually a clear distinction between them and in how they have progressed throughout history (McClellan and Dorn 2006).

Practical knowledge is a key component within technology. Although knowledge is a prerequisite for developing technologies, it is not necessary for that knowledge to be either scientific or systematic. In fact, there is a clear distinction between the terms "knowledge," "science," and "technology." "Knowledge" is a state or fact of knowing. The *American Heritage Dictionary* defines the term as "familiarity, awareness, or understanding gained through experience or study." (4th edition, 2004). As stated earlier, "technology" is the use of that knowledge to solve practical problems. Like technology, "science" also uses knowledge but in a systematic manner. Accordingly, "science" is defined as the systematic knowledge of the physical or material world gained through observation and experimentation. In short, science is an organized knowledge, and technology is a practical application of that knowledge (Spencer 1897; Francis 2007).

The history of architectural technology follows more or less the same path as the overall history of technology, but with additional elements. Knowledge, technology, and science are incorporated with art and materials to define "architectural technology." However, as in other fields, these terms have different meanings when applied to architecture. Parallel to the original definition of "technology," this book defines "green technology" as: (1) sustainable, ecological, and performative methods and tools; (2) an efficient means to an end; and/or (3) as an ensemble of green buildings.

But green technology also needs green architectural practices, like the creation, fabrication, and the use of green concepts. This includes nonmaterial technofacts, such as green energy generation and energy retention methods (e.g., passive design methods). Although, they require different materials to function, architectural green technologies are *not* architectural green materials. For instance, the act of converting solar energy into electrical energy is a technology, not a material.

This book investigates and surveys green technologies that may be organized into two broad categories: (1) energy generation and (2) energy retention. Energy generation looks at alternative green energy resources, which are ecological, sustainable, and healthy for the environment and for people. Energy retention tackles one of the oldest architectural problems in history, but which is still of paramount importance for green architecture today.

Energy Generation

In brief, energy generation depends on resources capable of producing energy and related extraction methods necessary to utilize these resources. Energy generation is directly related to human needs, and energy consumption can be correlated with a society's economic growth and development (Chevalier 2007). Today, the amount of energy needed is greater than ever. Unfortunately, more than 80 percent of our energy generation today is through fossil-based resources, which are exhaustible and nonrenewable.

Therefore, it is daily becoming more critical to develop new energy generation resources, technologies, and methods. Interest in generating green energy, because of its sustainable and ecological nature, has increased dramatically within the last couple of years. Mandates for energy sustainability (recommended by federal and state governments) have particularly facilitated this increase. Although currently these mandates are voluntary, 27 states have issued renewable portfolio standards (RPS), and a 2009 legislation introduced by Senator Tom Udall requires utility companies to produce at least 25 percent of their electricity from renewable resources by 2025.

The following energy resources do not depend on fossil-based resources, sustain the environment rather than deplete it, and offer great promise for architecture:

- Bioenergy
- Solar energy
- Geothermal energy
- Wind energy
- Hydro energy
- Blue energy
- Fuel cell energy
- Hybrid systems energy

Bioenergy

Bioenergy is energy derived from biomass, which is defined as all living plant matter and organic waste. Examples of these are: forestry residues, trees, grasses, animal and ethanol waste, sewage, garbage, wood construction residues, landfill gas, and other components of municipal solid waste (Tester, Drake, Golay, Driscoll and Peters 2005). Biomass is a renewable resource, meaning it is a part of the flow of resources that occurs naturally and repeatedly within the environment.

Bioenergy needs a continuous carbon cycle between the atmosphere and the earth. In this cycle, carbon dioxide is taken from the atmosphere for plant processes such as photosynthesis and converted into biomass. From this process, additional CO_2 is produced and transformed into energy (see Fig. 4.1). In order for this process to be effective, the continuity of the carbon cycle is essential. This explains why fossil-based energy resources are not considered biomass; their ancient biological origins have been out of the carbon cycle for too long.

Throughout history, biomass had been the primary fuel source for all civilizations. But as the Industrial Revolution progressed, especially in Europe, forests were severely depleted and coal was gradually introduced as a replacement fuel. In the United States, biomass was a primary source of energy up until the nineteenth century. However, by 1885 its use was being outpaced by coal, and by 1915, by oil and gas (Hottel and Howard 1971; Tester, Drake, Golay, Driscoll and Peters 2005). Today, many developing countries still use biomass for 90 percent of their energy source, especially for daily activities such as heating and cooking. Although biomass is an excellent green energy source, the current method of collection, production, and processing creates harmful effects. Primitive processing techniques and inadequate devices for energy conversion waste most of the biomass and create unhealthy living environments for people (e.g., massive indoor pollution). In addition, uncontrolled overpopulation places an overwhelming demand on indigenous biomass, causing land depletion and even desertification, as seen in some parts of Africa today (Hottel and Howard 1971).

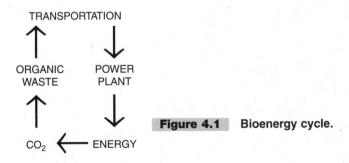

Figure 4.1   Bioenergy cycle.

BIOENERGY TYPES

There are primarily two types of bioenergy—traditional and advanced. *Traditional* bioenergy comes mostly from solid biomass sources, such as wood, charcoal, and other biomass pellets. Currently, over 80 percent of the energy from traditional solid biomass sources consumed as fuel for heating and cooking is generated with minimal efficiency (DOE 2006). *Advanced* bioenergy requires converting biomass into a liquid and/or gas form in order to produce electricity. Advanced bioenergies, such as biogas and liquid biofuel, have increased dramatically in recent years; other forms (e.g., ethanol, biodiesel, and algae fuel) are in use but are still developing technologies.

Solid bioenergy Solid bioenergy is derived from solid renewable resources. There are many forms of solid bioenergy material that can be directly used in gasification and combustion technologies. Agricultural and forestry biomass (and their by-products) constitute most of the solid bioenergy raw materials (see Fig. 4.2). They include:

- Forestry
  - Wood pellets
  - Woodchips

Figure 4.2   Wood pellets made from compacted sawdust have high combustion efficiency and are used as biofuel. (*Courtesy of Scion.*)

- Charcoal
- Sawdust
- Agriculture
  - Straw
  - Husks
  - Stalks
  - Bagasse (fibers left after sugarcane or sorghum stalks are crushed in juice extraction)
- Other
  - Construction waste
  - Municipal by-products

Liquid bioenergy   Liquid bioenergy is derived from plants and animal fats. There are two major liquid bioenergy types—bioethanol and biodiesel. Although they only make up about 2 percent of the transportation fuel today, they are expected to replace existing fossil fuels in the future.

- Bioethanol
  - Made from sugar and starches

- Biodiesel
  - Made from plant oils and animal fats

Gas bioenergy   Gas bioenergy is derived from methane and carbon dioxide, which are produced when bacteria break down biomass (e.g., animal, municipal and landfill waste, and energy crops). It can be mixed with (or used as an alternative for) natural gas. Major forms of gas bioenergy are

- Biogas
- Biopropane
- Syngas
- Synthetic natural gas (SNG)

Advantages of bioenergy
- Directly extracted from biomass
- Renewability and domestic availability
- Evenly distributed energy source
- Biomass utilization diverts landfill accumulation
- Potential to prevent CO_2 buildup in the atmosphere

Disadvantages of bioenergy
- Low energy density (compared to coal, liquid petroleum, or other petroleum-derived fuels)
- May contribute to environmental pollution

- Contributes to the depletion of earth's resources (land, water, plants, forest, food)
- High embodied energy (transportation costs do not justify the energy savings)

BIOMASS CONVERSION PROCESSES

To generate clean bioenergy requires biomass power generation facilities with advanced technologies that can process these materials under controlled conditions. Bioenergy can be generated by thermochemical, biochemical, and combustion processes (Moo-Young, Lamptey, Glick and Bungay 1987; Wereko-Brobby and Hagan 1996; Klass 1998; Song, Gaffney and Fujimoto 2002).

- **Thermochemical:** Although this process does not produce bioenergy, it converts biomass into a convenient form of bioenergy carrier, such as methanol, gas, or oil. This conversion is accomplished by carbonization, gasification, or pyrolysis. For more information about the thermochemical process, see *Progress in Thermochemical Biomass Conversion* (Bridgwater 2001), *Pyrolysis and Gasification of Biomass and Waste* (Bridgwater 2003), and *Advances in Thermochemical Biomass Conversion* (Bridgwater 2008).
- **Biochemical:** This process uses microorganisms to produce bioenergy. The biochemical production process uses several different technologies, such as (1) anaerobic fermentations that convert plant, human, and animal waste into biogas and fertilizers; and (2) anaerobic digestion that uses microbial decomposition of biomass in landfills. For more information about the biochemical process, see *Biotechnology of Biomass Conversion: Fuels and Chemicals from Renewable Resources* (Wayman and Parekh 1991), and *Biomass Conversion and Technology* (Wereko-Brobby and Hagan 1996).
- **Combustion:** This process is used for direct power generation. It is one of the most efficient conversion technologies today with 30 percent efficiency in electricity generation and more than 75 percent in cogeneration of electricity and heat. For more information about the combustion process, see *The Handbook of Biomass Combustion and Co-firing* (Loo and Koppejan 2008), and *Biomass Combustion Systems* (Reupke, Sarwar and Tariq 1994).

Solar Energy

The second type of green energy generation is solar energy. Solar energy is gained directly from the sun, our most abundant source of energy. The sun emits radiation during the fusion process, which produces various wavelengths of electromagnetic radiation (see Fig. 4.3). The earth captures only a small fraction of that energy from the interstellar void (Kambezidis and Gueymard 2004; Vita-Finzi 2008), and about 30 percent of the radiation that does reach earth is immediately reflected by various layers (i.e., atmosphere, clouds, the earth's surface) (Houghton, Ding, Griggs, Noguer, Linden and Xiaosu 2001).

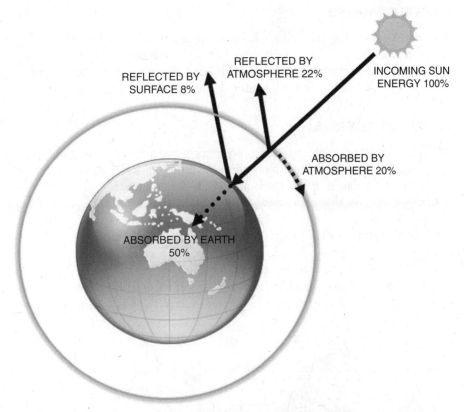

REFLECTED BY
ATMOSPHERE 22%

REFLECTED BY
SURFACE 8%

INCOMING SUN
ENERGY 100%

ABSORBED BY
ATMOSPHERE 20%

ABSORBED BY EARTH
50%

Figure 4.3 **Solar radiation distribution. The earth and the atmosphere absorb approximately 70 percent of the solar radiation. Thirty percent of the incoming radiation is reflected.**

Consequently, the amount of solar energy the earth receives in a year is approximately 3.1 million exajoules (EJ) (Houghton, Ding, Griggs, Noguer, Linden and Xiaosu 2001). According to the 2008 International Energy Outlook Report, our global energy consumption is 462 EJ/year (2005) and is expected to double by the year 2030 (DOE 2008). The vast discrepancy between these production and consumption statistics clearly show that the earth receives approximately 6900 times more energy from the sun than we consume globally. If harvested correctly, even a small fraction of the sun's tremendous energy output could provide more than we need.

Existing solar technologies convert solar energy into other forms of energy, namely electricity and heat. Although there are numerous ways to convert solar energy into usable energy (Sark, Patel, Faaij and Hoogwijk 2006), three primary conversion technologies are used:

1 Greenhouse
2 Solar thermal
3 Solar electricity

GREENHOUSE

The passive greenhouse technology is the oldest and simplest way to harvest solar energy, wherein the heat of the sun penetrates a special window system and is trapped inside. This system has been primarily used to provide adequate temperature control for plants in cold weather and climates (McCullagh 1978; Hanan 1997; Marshall 2006).

SOLAR THERMAL

Solar thermal technology uses the same principle as the greenhouse technique (i.e., using sunlight to create heat) but is more advanced and needs water to operate. It was originally developed to pump water in the nineteenth century. Once solar thermal technologies receive sunlight, they concentrate the light and generate heat. The generated heat warms the water, which can either be circulated and used directly, or can be used to drive a turbine that generates electricity (Peuser, Remmers and Schnauss 2002). There are three types of solar thermal systems: parabolic troughs, parabolic dishes, and power towers.

The parabolic trough is the most established low-cost solar thermal technology available today (see Fig. 4.4). Though parabolic dish systems have great potential for performance and cost efficiency, they are currently more expensive than the trough

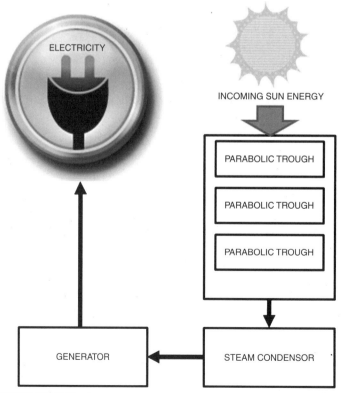

Figure 4.4   **Parabolic trough collectors.**

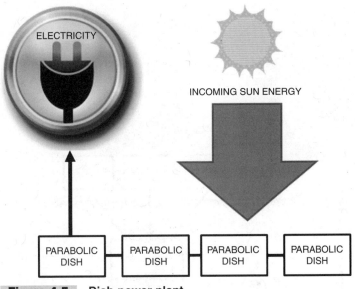

Figure 4.5   **Dish power plant.**

systems and are still under development (see Fig. 4.5). One of the main advantages of parabolic dish systems, however, is their flexibility. Since these systems are modular, they can be used independently as a single dish or grouped together to create solar thermal dish farms. Power towers are mostly used for large grid systems in the 50 to 200 MW size, and require a large area to operate (Peuser, Remmers and Schnauss 2002; Duffie and Beckman 2006) (see Fig. 4.6). They work with an extensive number of light-tracking mirrors (heliostats) that reflect light to the tower receiver.

SOLAR ELECTRIC (PHOTOVOLTAIC)

Of the solar electric systems currently available, photovoltaic technology is the most advanced and mature. The material used in this technology is covered in Chapter 7, "Smart Materials," but briefly, photovoltaic technologies are nonmechanical devices that convert sunlight into electricity. First discovered by French physicist Edmond Becquerel in 1890, and later developed by Bell Laboratories researchers in 1954, photovoltaic technologies are widely used in our everyday lives. Their applications range from pocket solar calculators to large systems that power our buildings.

A photovoltaic system consists of: photovoltaic cells, mounting hardware, electrical connections, power conditioning equipment, and an energy storage device (see Fig. 4.7). Individual *photovoltaic cells*, which can be as small as a dime, can generate between 0.5 to 1.2 V of electricity. Individual cells can be grouped into *modules* to form larger collectors, which can in turn be further grouped into *photovoltaic arrays,* necessary for industrial-level electricity production. The number and size of the modules can vary depending on: the availability and intensity of the sunlight, the geographical location of the modules, and the user's needs.

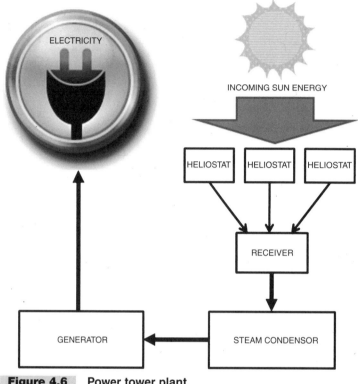

Figure 4.6   Power tower plant.

A basic photovoltaic (or solar) cell typically produces only a small amount of power. To produce more power, cells can be interconnected to form modules, which can, in turn, be connected into arrays to produce even more power. A majority of photovoltaic cells are made of silicon, and there are three main photovoltaic technologies used commercially in the market today: (1) single crystalline, (2) polycrystalline, and (3) thin film.

Single crystalline photovoltaic cells   Single crystalline photovoltaic (SCPV) is a first-generation photovoltaic cell, and while it is currently the most efficient technology, it also has high production costs. SCPVs are made of large cylindrical, single-crystal silicon, which is produced in an oven and sliced into wafers. This is a very clean and efficient process with very low degradation (between 0.25 and 0.50 percent per year). SCPV's electricity conversion efficiency rate averages between 12 and 15 percent, and it has an exceptional 24 percent efficiency in laboratory conditions.

Polycrystalline photovoltaic cells   Polycrystalline photovoltaic (PCPV) cells are made of silicon that is cast into cylinders and then sliced into wafers. Since this process is less precise than SCPV fabrication, it has lower manufacturing costs. PCPV's conversion efficiency is slightly lower (10–11 percent on average) than SCPV, but overall it is quite comparable. Degradation, assembly, and doping processes are the same as for SCPV.

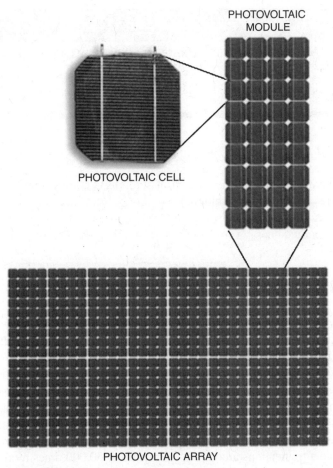

PHOTOVOLTAIC
MODULE

PHOTOVOLTAIC CELL

PHOTOVOLTAIC ARRAY

Figure 4.7   **Photovoltaic systems.**

Thin film   Thin film is a low-cost, low-efficiency photovoltaic technology with an average electricity conversion rate of 5 to 7 percent. Its manufacturing process is different from that of crystalline systems, as it involves vaporizing and depositing silicon onto glass, stainless steel, or plastic (see Table 4.1). It is also less expensive to manufacture because of its large module production, use of fewer semiconductor materials, and lack of individual wiring and framing costs. Because of its low efficiency, however, thin film technology is primarily used in small consumer products. Recently, more advanced, efficient (13.5 percent conversion rate) thin film materials have been created by stacking multiple layers of photovoltaic materials within a module.

■ Free energy with unlimited supplies
■ Environmentally safe (produces no pollution)

TABLE 4.1 EFFICIENCY RATE OF PHOTOVOLTAIC TECHNOLOGIES			
	SINGLE CRYSTALLINE	POLYCRYSTALLINE	THIN FILM
Typical efficiency	12–15%	11–14%	5–7%
Maximum efficiency	24.7%	19.8%	13.5%

Disadvantages of solar energy

- High initial cost
- Inconsistent delivery of energy (the amount of sunlight is not constant, depending on location, time, and weather conditions)
- Large surface area installation is required to produce sufficient energy
- Indirect high embodied energy (production of solar energy technologies)

Geothermal Energy

A third method of "green" generation of energy is geothermal. Geothermal energy is produced by using the heat below the earth's surface. This heat originates from the earth's core 4000 miles below the surface, where temperatures can reach up to 9000°F. Although most geothermal basins are deeply rooted underground and are, therefore, unseen, they occasionally burst forth in the form of hot springs, geysers, volcanoes, and fumaroles. In addition to the earth's core, geothermal energy may emerge from other sources, such as from continental plate frictions and the decay of naturally occurring radioactive substances within the crust. Geothermal resources have the potential to provide a tremendous amount of energy, up to 50,000 times more energy than all the earth's oil and gas resources combined (Berinstein 2001).

There are various ways to extract geothermal energy. The simplest is to drill into geothermal reservoirs in order to bring the heat source (steam or hot water) to the surface. Geothermal heat pumps are necessary for residential use, and, for commercial use, power plants are built on the surface to convert geothermal energy into electricity. There are three main types of power plants: (1) dry steam, (2) flash steam, and (3) binary power.

Dry steam plants Dry steam plants (DSP) are the most widely used geothermal energy power plant today. They must be located near accessible steam reservoirs, where the steam is tapped and piped directly to the plant in order to power the generator's turbines.

Flash steam plants Flash steam plants (FSP) convert high pressure hot water into steam, which powers the generator's turbines. Their efficiency is 50 percent lower than for DSPs because of energy losses during the water to steam conversion (Berinstein 2001). However, the condensed water can be recycled and reused.

Binary power plants   Binary power plants (BPP) are complex versions of flash steam plants and are especially useful in that they allow cooler geothermal reservoirs to be tapped. The cooler reservoir water is pumped into a heat exchanger and then back down into the reservoir. Then another liquid with a lower boiling point is rapidly pumped in. The heat is sufficient to vaporize the second liquid, and the steam that is produced powers the turbines. As in FSPs, the second liquid is condensed and reused.

GEOTHERMAL HEAT PUMPS

Geothermal heat pumps (GHP) are used to heat or cool residences, rather than relying on fossil fuels. They run on the same principle as commercial power plants, utilizing the constant heat source of the earth's interior. The Environmental Protection Agency (EPA) states that a geothermal system can save up to 30 to 70 percent on home heating and 20 to 50 percent on home cooling over conventional systems (EPA 2008). However, installing these systems is still quite expensive.

A geothermal heat exchanger system consists of indoor pump equipment, a ground loop, and a flow center to connect the indoor and outdoor equipment. The ground loop uses the temperature of the earth or water, several feet underground, to heat or cool the dwelling. A pump circulates a temperature-sensitive fluid through this ground loop, and since the loop is buried below the freezing line, the temperature stays constant all the time (approximately 60°F). GHPs transfer stable temperatures into houses but in reverse order according to the season. As such, in the winter warm fluid carries heat into the house, and in the summer cool fluid draws heat out of the house.

Advantages of geothermal energy
- Clean, no polluting emissions
- Reliable
- Flexible
- Regional (contributes to local economy)
- Resource availability

Disadvantages of geothermal energy
- Not a renewable resource like sunlight and wind
- Location-specific
- Problems with accessibility
- Potential environmental damage (e.g., erosion, sedimentation, toxic antifreeze solutions in heat pump systems)
- Residential heat pump systems are expensive

Wind Energy

Wind energy is yet a fourth form of green energy, related to solar energy, as wind is generated by solar patterns and their influence on the earth's topography. The planet's rotation, climate, and topography contribute to the speed and direction of the wind that will be harnessed. However, wind energy provides a significantly smaller amount of energy than solar energy. The global theoretical wind energy potential is only 2 percent of the amount of solar energy that reaches our atmosphere (Hubert 1971; Sark, Patel, Faaij and Hoogwijk 2006).

Wind energy turns a windmill's blades on a rotor that is connected to a main shaft. The main shaft spins a generator, producing energy. The amount of energy generated depends on various factors, such as the speed and the direction of the wind. Even though strong winds can produce more energy, it is difficult to design and maintain windmills that can withstand such force. Another problem is proximity of the wind generation facilities to the distribution centers and/or to the homes. The farther the distance, the more the distribution lines that need to be extended. This affects the quality and cost of the energy transmission.

There are three types of wind machines used today: (1) horizontal axis, (2) vertical axis, and (3) wind amplified rotor platform.

Horizontal axis   The horizontal axis is the most widely used wind machine today. This type has three long blades similar to an airplane's propeller, installed at the top of a tower whose height is comparable to that of a 15- to 20-story building. The taller and wider the wind machine's blades are, the more wind that can be captured.

Vertical axis   Vertical-axis machines have vertical blades attached from top to bottom. These machines are thought to be more efficient than horizontal-axis types, generating more wind energy with less wind. However, they account for a very small percentage of the windmill power generation today.

Wind amplified rotor platform   Wind amplified rotor platform (WARP) is a completely different wind machine design, as it does not have blades. Instead, independent modules are stacked on top of each other with small high-capacity turbines mounted on each module. Their concave surfaces significantly amplify the wind's speed. WARP systems are considered to be very efficient, using less wind and land area, yet generating more power. These systems are currently under development and are expected to be used in offshore oil platforms and remote telecommunication towers.

Advantages of wind energy
- Clean, no emissions
- Reliable
- Renewable
- Regional (contributes to local economy)
- Resource availability

Disadvantages of wind energy
- Location-specific
- Problems with accessibility

- Requires land allocation and wind farming
- Noisy
- Potentially dangerous for birds

Hydro Energy

Hydro (water) energy, generated from the force of moving water, is a fifth ecologically sensitive type of generating power. It is a clean, renewable energy resource but usually requires large installations, such as dammed water, to drive a water turbine and generator. The amount of generated energy is controlled by pumping water to different reservoir levels. This pumped storage technique is used in large-scale grid power stations, and today they provide 20 percent of the world's energy and 52 percent of the total electrical energy generated from renewable sources (REN21 2009) (see Fig. 4.8).

Although hydro energy is primarily generated by large installations, there are smaller, mobile hydro-generators available for individual home use. These devices, which have a generating capacity between 50 and 1000 kW, produce enough electricity to power homes built near water or for home-boats. The advantages of the mobile hydropower generators are that they operate on any available stream without requiring additional water storage, reservoir, and/or dams. The amount of energy generated depends on the amount of available water, the power of the stream, and the size of the residence. The problems with small hydro-generators are their complexity and site

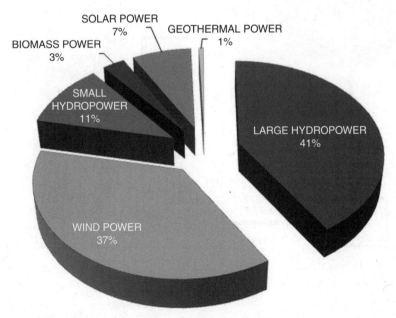

Figure 4.8   Renewable energy added to the existing capacities, 2008 (estimated). *(Source: REN21 2009)*

specificity. In addition, while these devices are inexpensive to operate, they are costly to buy, install, and maintain.

Blue Energy

Yet another source of clean energy is blue energy, also called *osmotic energy*. It is generated from a chemical reaction between fresh water and sea water. Energy is retrieved from a dilution process, designed to balance the salt concentration differences between the two solutions, in a process called "pressure osmosis" or "reverse electrodialysis" (see Fig. 4.9).

This sixth source is a new, promising, renewable clean energy source with no harmful environmental effects. It can either be installed near a salt water resource (i.e., at the mouth of a river) or operated independently using stored water. Although the technology has been developed, there are no commercial applications available today. This is primarily because its complexity requires operational expertise to run, and the costs associated with its installment and use are high.

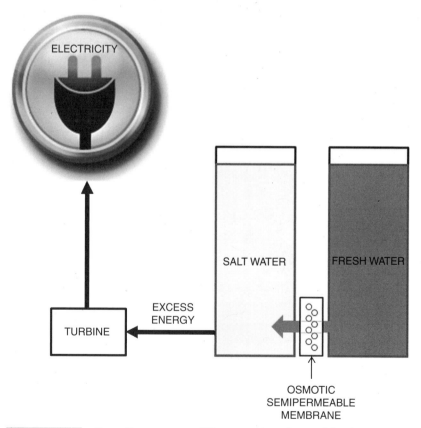

Figure 4.9 Osmotic pressure differences in salty and fresh water creates blue energy.

Fuel Cell Energy

Fuel cell energy is a seventh promising method of generating energy. A fuel cell is a kind of battery that produces electricity from the reaction between an externally supplied fuel and an oxidant, in the presence of an electrolyte. There are different fuel cell combinations, depending on the type of fuel and/or oxidant. For example, a hydrogen fuel cell uses hydrogen as the fuel and oxygen as its oxidant. Other fuels may include (but are not limited to) methane, ethanol, biofuels, or chlorine. Although they resemble batteries, these devices are different in that they generate but do not store energy. Traditional batteries store energy chemically in a defined and enclosed system, whereas fuel cells consume renewable reactants. Thus, these cells cannot store energy, but they can provide nonstop, continuous operation as long as the required energy flow is maintained.

Fuel cells are manufactured in different sizes and capacities, and can be used for products ranging from small consumer electronics to energy generators for large buildings. For home use, fuel cells are currently being developed by various manufacturers, such as General Electric Fuel Cell Home Power Plant (HomoGen 7000), Tokyo Gas with Ballard Power Systems and Matsushita Electric (1 kW combined heat and power fuel cell generator), and Astris Energy of Canada with Alternate Energy Corp (4-kW residential fuel cell).

Fuel cells provide clean, renewable energy, and do not need distribution lines (see Fig. 4.10). Small-size home fuel cell units generate about 10 kW and produce heat as a by-product. This heat can then be used as a thermal cogeneration system for applications such as domestic hot water and space heating. Fuel cells can be used in conjunction with power grid systems, or be used independently as an off-the-grid system

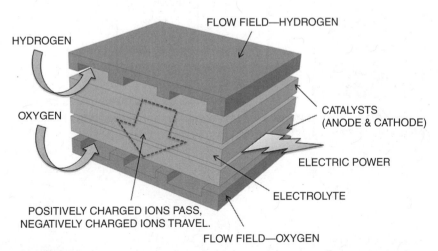

Figure 4.10 Fuel cells provide clean, renewable energy without distribution lines.

in remote areas. The main problems associated with using this energy for home use are its complexity, fuel regulations and liability issues, and high cost.

Hybrid Systems

The eighth and last source of energy generation with no damaging environmental impact is the hybrid system. This is a method that uses two or more distinct power sources to run a device. Examples could include: an on-board rechargeable energy storage system (RESS) with a fueled power source (internal combustion engine or fuel cell); air and internal combustion engines; and photovoltaic modules and wind turbines with electric power (see Fig. 4.11).

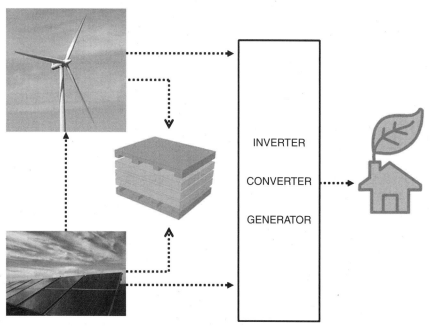

Figure 4.11   Hybrid power systems combine multiple sources to deliver continuous electric power.

5

GREEN TECHNOLOGIES:
ENERGY RETENTION

We will turn our attention now from the *generation* of energy to the *retention* of energy. Although the installation of an energy system in a home represents 30 percent of the cost, energy retention efforts account for more than 75 percent of the energy expended in the average household. The amount of energy that can be retained, rather than dissipated, depends on several factors, many of which are architectural. These include the size, shape, style, orientation, and construction of the building, as well as the insulation, heating/cooling, and envelope systems within the building. Nonarchitectural factors include the location, local climate, and users' preferences.

Insulation Systems

Thermal building insulation is essential for energy retention, which (as noted) is one of the most important issues and oldest problems in the architectural field. Insulation fortifies the building against temperature fluctuations in the environment, reduces unwanted cold or heat, and affects an occupant's budget significantly (positively or negatively). Up to 40 percent of the energy used to heat or cool a building is caused by air leakages, which continually drain additional energy in order to maintain a constant ideal indoor temperature. However, as much as 30 percent of this energy expenditure can be eliminated by proper insulation. Green insulation systems are adaptations of traditional systems, offering enhanced performance with the added benefit of being a sustainable and ecological solution to energy retention problems. The primary green insulation systems are: (1) bio-based insulation, (2) rigid panel insulation, (3) glass-based insulation, (4) wood-cement forms, (5) natural fiber insulation, and (6) radiant barriers.

BIO-BASED INSULATION

Bio-based insulations are made of renewable biomaterials that are more energy efficient, healthier, and more durable than traditional thermal insulation systems (see Fig. 5.1). They can be applied as foams, spray-foams, polymers, or biofibers, and since many of these bio-based systems are water-blown and soy-based, they are clean and safe for the environment.

Soy-based insulation is made of renewable soy beans. When applied as spray-foam insulation, it can expand to 100 times its volume, easily insulating and sealing voids and air gaps even after one application. Bio-based insulations are: lightweight; contain no volatile organic compounds (VOCs), chlorofluorocarbons (CFCs), or hydrochlorofluorocarbons (HCFCs); provide a minimum R-value of 3.5 per inch that stays the same over time; have excellent thermal and acoustical properties; and more importantly, meet U.S. government requirements for renewable resources. Moreover, they have a class 1 fire rating, have low levels of free-floating dust and allergens, and are not affected by moisture, mold, or insects.

RIGID PANEL INSULATION

Rigid panel insulation (RPI) systems are one of the most widely used insulation systems today. Because of their firmness and rigidity, RPIs are mostly used as roof and wall coverings, as they contribute to the building's overall structural strength. They can be applied as part of the roof assembly, or affixed directly to the interior or exterior part of the wall. Rigid foam panels can be made fireproof when shielded with additional fire-resistant materials (e.g., wallboard), contain no CFCs, and have significantly high R-values (the unit of measurement gauging the degree of protection against heat gain or loss).

RPIs can be made from various types of base materials, including: cellular glass, wood fiber, recycled beads, expanded polystyrene (EPS), extruded polystyrene (XPS), polyisocyanurate (ISO), perlite, gypsum board, and composites.

FEATURES	TRADITIONAL	BIO-BASED
Contains HFCs, VOC, PBDEs		✓
No shrinkage over time		✓
Air barrier		✓
No food source for insects	✓	✓
No support for mold/mildew		✓
Made from renewable source		✓
Performance stability		✓

Figure 5.1   Bio-based insulation systems are clean, durable, and energy efficient alternatives to traditional insulation systems.

GLASS-BASED INSULATION

Glass-based insulation systems are made of glass fibers combined with reinforcing agents. There are different forms of glass-based insulation systems. Fiberglass is the most widely used glass-based insulation system today. Recycled glass insulation is a composite made from glass fibers and a polymer, and cellular glass is made of crushed glass mixed with foaming materials. Glass-based insulation systems are used for roofs, attics, walls, ducts, pipes, as well as other home appliances and equipment.

Glass-based insulation products (i.e., fiberglass batts and rolls) have relatively small deterioration rates (1–3 percent) and high R-values, ranging from R-11 to R-38. They can be blown in any void to any desired R-value. Since the base material is made of sand and recycled glass, it is naturally noncombustible, requiring no additional chemical treatment. Moreover, glass-based insulations are not absorbent and retain their R-value, even after they are exposed to moisture.

INSULATING WOOD-CEMENT FORMS

Insulating wood-cement forms (IWCF) are stay-in-place systems, which can be cast in place or installed as preassembled pieces that lock together to create structurally solid, insulated forms (see Fig. 5.2). IWCFs can be made from a variety of different materials, such as cement-bonded wood fiber, polystyrene beads, and expanded polystyrene (EPS).

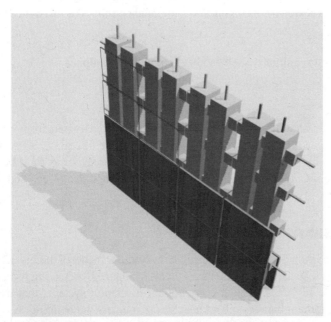

Figure 5.2 Insulating wood-cement forms are installed as preassembled pieces that lock together to create structurally solid, insulated forms.

IWCFs are highly efficient insulation systems, especially for low-rise residential and commercial constructions, as their structure minimizes air leaks and heat loss. In addition, they are also durable insulation systems with a high sound absorption rate, a long life span, and relatively low maintenance costs.

NATURAL FIBER INSULATION

Natural fibers are made from animal, plant, and mineral resources. They contain no VOC or chemical irritants, have excellent thermal and sound blockage, and have a very low embodied energy. Natural fibers inherently have the ability to absorb moisture, and can breathe and react to climatic changes. Overall, natural fiber insulation is a healthier, more environmentally friendly alternative to synthetic fiber insulation. Natural fibers are classified based on their origins.

Plant sources

- *Seed fiber.* Extracted from seeds or seed cases, such as hemp, kapok, and cotton
- *Leaf fiber.* Extracted from the leaves of plants, such as sisal and abaca
- *Wood fibers.* Made directly from wood or by mixing wood with other plant fibers. Since they are not inherently moisture resistant, wood fibers must be coated with enhancers or other adhesive materials.
- *Cellulose fiber.* Made from 80 percent recycled postconsumer paper waste. Cellulose fiber insulation is applicable in loose-fill form only. Typically it is blown or poured into hard-to-reach areas, such as attics and finished wall cavities. It, too, can be sprayed with water-based adhesives to enhance moisture resistance.

Animal sources

- *Hair.* Extracted from animals, such as sheep, mohair, and alpaca
- *Secretions.* Fiber collected from the cocoons of insects, such as silk worms

Mineral sources

- These either occur naturally or are modified from fibers extracted from minerals. Their visual and performative properties closely resemble fiberglass. They are primarily applied as loose-fill and batting in hard-to-reach areas, or as rigid boards in roofs and attics. Since they are highly fire-resistant, they are also often used for furnace and chimney insulation.

RADIANT BARRIERS

Radiant barriers are composite material insulation systems made of highly reflective materials or coatings that stabilize a structure's thermal balance by reducing its heat gain and/or heat loss (see Fig. 5.3). Unlike conventional insulation systems that trap heat within the material, radiant barriers reflect heat before it enters the building. As they do not store and then reemit the heat inside a building, radiant barriers are more energy and cost-efficient than conventional systems. Radiant barriers are often applied in attics or underneath the roofing layers in residential, commercial, and industrial buildings.

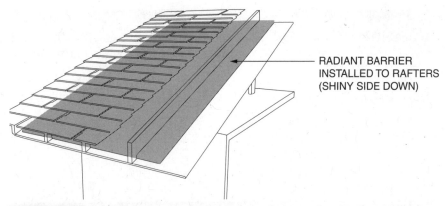

Figure 5.3   **Radiant barriers prevent over 95 percent of the attic heat gain and reduce the temperature by 30 percent.**

Heating/Cooling Systems

Heating and cooling expenditures account for more than 45 percent of the total energy consumption in residential buildings (see Fig. 5.4 and Table 5.1).

Regardless of the energy source used for heating/cooling, buildings currently use either central or local systems to adjust their thermal balance. Central heating systems are used in mass dwellings in cold climates to simultaneously distribute a treated thermal substance (hot/cold, air/fluid) to multiple spaces. However, delivery methods vary depending on the distribution system. For example, forced air is used for heated air

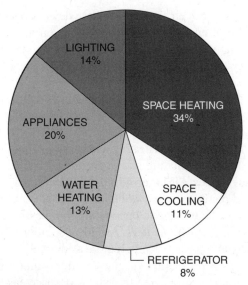

Figure 5.4   **Total energy consumption in residential buildings.**

TABLE 5.1   2005 RESIDENTIAL ENERGY EXPENDITURE BY ENERGY TYPE*

	NATURAL GAS	PETROLEUM	COAL	ELECTRICITY	TOTAL	%
Space heating	43.7	18.3	0.03	13.6	75.63	39.1%
Space cooling	0	0	0	23.2	23.2	12.0%
Water heating	14.1	2.6	0	11.5	28.2	14.6%
Lighting				20.8	20.8	10.7%
Refrigeration				14.2	14.2	7.3%
Appliances				22.4	22.4	11.6%
Other	0.5	3.3		5.3	9.1	4.7%
Total	58.3	24.2	0.03	111	193.53	100%

Sources: EIA, Annual Energy Outlook, Table A2 and Table A4, (Feb. 2007).
*The unit is $ Billion.

systems, whereas for fluid and electrical systems, convection and radiation are used to transfer heat. Central heating systems use a furnace, boiler, or heat pump to transfer heated air, water, or steam through ductwork or radiators. For smaller environments and/or temporary circumstances, heating may also be provided via local, non-mechanical means, such as electric space heaters, baseboard and wall heaters, heat pumps, and passive systems.

Some of the primary delivery devices are

- *Furnaces.* Used to heat air only (i.e., hot air furnaces). These devices operate efficiently and can use a wide range of energy resources (e.g., natural gas, propane, biomass, fossil oil, electricity). Natural gas furnaces with condensing designs provide 20 percent higher efficiency than Annual Fuel Utilization Efficiency (AFUE) standards (95 vs. 78 percent, respectively).
- *Boilers.* Used to heat water for domestic use and hydronic heating systems. Similar to furnaces, boilers also run on a variety of energy resources, including electric and biomass.
- *Heat pumps.* Used to heat air and water for both heating and cooling. These systems collect heat from a variety of sources (including air, water, or specialized liquids) and circulate the substance through pipes. Air-based heat pumps are used when cooling a structure, and ground-based pumps are used when heating one. Both use electricity to pump the substance (air or liquid) and to operate the compressor.
- *Electric space heaters, baseboard, and wall heaters.* Are ideal solutions for small spaces and local heating requirements, as they do not require a comprehensive ductwork and/or piping. Wall heaters are installed on the exterior wall, baseboard heaters are installed at the bottom of any interior wall, and space heaters are portable.

- *Radiant systems.* Heating cables or tubes are embedded within the construction material (i.e., concrete, tile, and wood) for dispersion of heat throughout a space. The same tubes can be used for cooling in summer as well. For cooling systems, tubing is mounted on the ceiling and then cooled water (mixed with additional additives such as glycol) is pumped through the tubes by a mechanical pumping device or a geothermal system. The advantages of these systems are their efficiency, cost, and their homogeneous transfer of temperature. The major disadvantages are the maintenance of the overall system, and the fact that the radiant cooling system is ineffective in humid environments. Therefore, it must be used in conjunction with a dehumidifier system.
- *Green roofs.* Offer a natural heating/cooling alternative by providing thermal insulation and evaporative cooling, due to water circulating from plants.

Ventilation Systems

Ventilation is essential to maintain healthy indoor air quality, as it circulates air within the building, as well as exchanges the inside air with the outside air. Ventilation systems control temperature and humidity levels, and remove airborne bacteria, odor, and dust.

There are two types of ventilation: (1) natural and (2) mechanical.

- *Natural ventilation* is the preferred ventilation method, as it uses operable windows and direct outside air circulation, when the temperature, wind, precipitation, humidity, and pollution levels are acceptable. If used correctly, natural ventilation can save a building up to 10 to 15 percent of its energy consumption. For example, in hot climates cool air should be circulated at night and then retained within the structure throughout the day by preventing additional air circulation. In addition to being desirable only when outdoor air quality permits, the amount of natural ventilation depends on the type, shape, placement, and size of the building and its openings. There are two primary natural ventilation methods: (1) cross-ventilation and (2) stack-ventilation. Cross-ventilation depends on wind-driven breezes, whereas stack-ventilation uses air density differences to create air movement across a building. Regardless of the type, natural ventilation entails certain building design requirements:
  - Buildings should be oriented with minimal obstruction to wind.
  - Natural ventilation is distributed better in narrow buildings and/or smaller-sized buildings with open plans.
  - Design must have an adequate number and size of openings.
  - Space planning must ensure minimal spatial obstructions, such as interior walls or partitions.
  - Buildings must provide operable windows, skylights, and attic ventilation. Windows should be placed in locations to maximize the wind and/or in opposite sites, as possible. When skylights and operable windows work in tandem, they act like a solar chimney, drawing heat out of the structure. Lastly, attic ventilation reduces heat transfer to other parts of the building.

■ *Mechanical ventilation* is a forced ventilation method that circulates the air, removes odors, and controls humidity within the building. Mechanical ventilation is often used in wet areas (e.g., food preparation, bathrooms) to control odor. Ceiling and window fans, and/or portable ventilation devices are used to circulate the air within the space. They cannot be used for air replacement unless a clear indoor/outdoor circulation pattern is established.

Water and Waste Management Systems

Buildings use one-sixth of the world's fresh water and contribute almost 22 percent to the waste stream (Roodman and Lenssen 1995). Therefore, for any building to be considered "green," water and waste management issues should be managed using sustainable, ecological, and high-performance methods and approaches.

A second aspect is solid waste, such as wood, concrete, glass, drywall, and asphalt shingles, which is the primary type of waste produced by buildings. According to a 2003 EPA study, the total estimated construction and demolition waste within the United States was 325 million tons.

WATER MANAGEMENT

Methods of conserving and collecting water
■ Use reduction:
  ■ Use EPA-approved water-efficient products, plumbing fixtures, and appliances.
  ■ Use sensors on public appliances to turn the water off when not used.
  ■ Repair all leaks.
  ■ Use weather/sensor-based irrigation controllers and/or water the lawn at the coolest time of the day (i.e., late evening and/or early morning).
■ *Gray water systems.* Gray water is collected wastewater from domestic processes (e.g., bathing, washing dishes, or laundry) that is reused for flushing toilets, watering landscapes, and irrigation. Gray water comprises three-quarter of domestic wastewater, and is different from the heavily polluted blackwater in terms of levels of biological contaminants and toxic chemicals (see Fig. 5.5).

Advantages
■ Applicable to any building
■ Reduces fresh water use
■ Helps groundwater recharge
■ Places less stress on septic tanks and treatment plants
■ Recycles wasted nutrients that can be used to nourish topsoil

Disadvantages
■ Requires startup costs, such as a pressurization pump, sump pump, and secondary plumbing
■ High maintenance

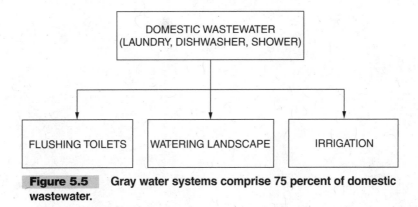

Figure 5.5   **Gray water systems comprise 75 percent of domestic wastewater.**

- Presents permit issues in some states for new construction.
- *Storm water systems.* Storm water systems are mostly used in commercial buildings as they solve watershed problems by capturing, moving, and treating the excess water off the building site. Drainage, flood control, and water treatment are the main principles for effective storm water management systems. Green buildings should use the following storm water solutions:
  - Bioretention: adds landscaping and vegetation to treat storm water runoff (e.g., rain gardens).
  - Harvesting: reduces the amount of water before it can cause flooding.
  - Infiltration: collects and filters storm water into an aquifer.
- *Composting toilets and waterless urinals.* In high traffic facilities and inadequate infrastructures, green buildings should use composting toilets and waterless urinals to eliminate water usage and sewage disposal (see Fig. 5.6). Waterless urinals use cartridge inserts filled with a sealant liquid, which collects and traps the urine without releasing the odor. Although the cartridge and the sealant are replaced periodically, a single waterless urinal saves 40,000 gallons of water per year.

Using little or no water, composting toilets are designed to convert human waste into compost via aerobic decomposition. These practical devices can be installed anywhere, from a remote cabin to a recreational vehicle. Composting toilets require that the solids be removed periodically, but the removal period varies depending on the size of the unit. Small units need to be emptied several times a year, while larger full-size units can last for several years (even decades) before the waste needs to be removed. With full size units, the waste volume decreases by 95 percent over 5 years, by which time the remaining waste has turned into mineralized soil that cannot be decomposed further.

WASTE MANAGEMENT

Waste management is the process of collecting, recycling, and disposing of waste materials produced by human activities. Waste management involves solid, liquid, and gaseous substances, some of which can be hazardous. As such, each requires a different method and procedure to process. Although it is a common practice to outsource

 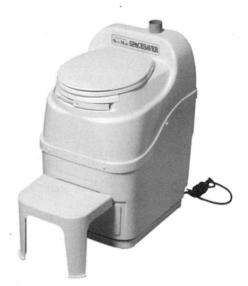

SLOAN WES-5000 WATERFREE URINAL. INSTEAD OF WATER, **WATERFREE URINALS** USE BIODEGRADABLE SEALANT LIQUID THAT FLOATS ABOVE THE URINE AND REMOVES SEWER GAS. INDIVIDUAL UNITS SAVE ANYWHERE BETWEEN 10,000 TO 40,000 GALLONS OF WATER URINAL/YEAR (COURTESY OF SLOAN)

SUN-MAR SPACESAVER **COMPOSTING TOILET.** SELF-CONTAINED COMPOSTING TOILET SYSTEMS FLUSH WASTE TO THE COMPOSTING UNIT BELOW THE TOILET. (COURTESY OF SUN-MAR)

Figure 5.6   Waterless urinals and composting toilets are effective methods of collecting and conserving water.

waste collection to commercial waste management companies, in most states local governments are responsible for the management of nonhazardous residential waste. Contractors and other private parties are responsible for disposing of nonhazardous commercial waste.

There are three widely used waste management concepts: (1) waste hierarchy, (2) extended producer responsibility, and (3) "polluter pays" principle.

Waste management concepts

■ *Waste hierarchy*. Commonly known as the 3Rs—"reduce," "reuse," and "recycle"— waste hierarchy is the most widely used waste management strategy. In actuality, the hierarchy has more than three steps, and some steps are "greener" than others (see Fig. 5.7). The overall objective of this concept is to (re) use waste by extracting extra (i.e., additional or new) value while generating a minimal amount of refuse.

■ *Extended producer responsibility*. Extended producer responsibility (EPR) is an integrated waste management strategy, which makes manufacturers accountable over a product's lifecycle (i.e., manufacture, assembly, recycling, and disposal). This strategy shifts the economic burden from taxpayers to manufacturers, who are responsible for including all of a product's environmental costs into the market price.

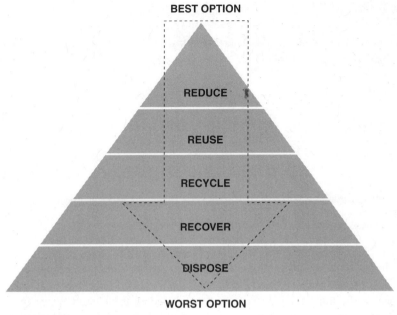

Figure 5.7   Waste hierarchy.

- *"Polluter pays" principle.* "Polluter pays" principle (3P) is an effective environmental policy tool that requires responsible party(ies) to pay for polluting the environment. Regardless of their intention or the cause of the pollution, the responsible party must pay for the costs associated with environmental cleanup, removal, and retributions to those who have been affected.

Waste management methods

Waste management is a complicated process, involving different methods that vary from disposal to recycling. While some traditional methods may release environmentally hazardous gases and pollutants, green waste management methods focus on sustainable and ecological solutions such as green disposal systems and recycling.

- *Green disposal systems.* Municipal waste disposal facilities process household, nonhazardous commercial, institutional and industrial solid waste, and construction and demolition debris. The most common disposal methods are landfills and incineration, but the use of anaerobic digestion as an ecological method has gained some momentum in the last decade.
- *Landfills.* Landfills are the most widely used waste disposal method in the world today. Although most third-world countries dump their refuse in waste land, technically landfills are different. Modern landfill facilities are highly engineered, regulated operations that are closely monitored for health and environmental protection.

TABLE 5.2 EPA'S ESTIMATES FOR THE OVERALL PERCENTAGE OF C&D DEBRIS	
Concrete and mixed rubble	40–50%
Wood	20–30%
Drywall	5–15%
Asphalt roofing	1–10%
Metals	1–5%
Bricks	1–5%
Plastics	1–5%
Source: Sandler, (2003)	

- Most of today's construction and demolition waste (C&D) ends up in either municipal or specifically devoted C&D landfill sites. According to the EPA, 136 million tons of building-related waste was generated in the United States in 1996. The majority of this waste was generated in building demolitions. Seven years later, the EPA estimated that construction and demolition waste was up to 325 million tons.

- Depending on the project from which it originated, C&D materials can vary greatly (see Table 5.2). For example, older structures often contain materials like lead piping and plaster, while plastics, laminates, and dry wall are typically found in newer structures.

- Advantages of landfills include saving on waste transportation costs and being able to process a variety of waste materials. One of the major problems with landfill waste decomposition is the production of methane, a noxious, hazardous greenhouse gas that directly contributes to global warming. In addition, once a site has been used as a landfill, it is considered polluted and cannot be reused for future developments.

- *Incineration.* Incineration (or combustion) is the process of burning waste in a controlled manner. Incineration facilities are capable of converting waste into energy, sorting chemical, biological, and physical compositions from the waste (including recyclables), and destroying harmful chemicals found within the waste. There are three types of incineration: (1) fixed-hearth (used mostly for medical and municipal waste); (2) rotary kiln (used mostly for industrial waste); and (3) liquid injection (used mostly for chemical industrial toxic waste).

- One of the major advantages of incineration is the permanent destruction of the waste; it does not require storing like landfills. Another advantage is the ability to generate energy from the waste incineration, such as steam for heating or electricity. The major concern regarding this method is its emission of toxic chemicals (e.g., dioxin), in spite of the modern environmental control systems.

- *Anaerobic digestion.* Anaerobic digestion is the process of decomposing waste in a closed oxygen-free environment. The waste is broken down into gaseous (methane) and solid by-products. Methane is burnt and then converted into electricity, and organic fertilizers are produced from the liquid and solid biodegradable wastes.

■ *Recycling.* Recycling is the process of extracting resources from waste, and/or adding additional use value to waste. There are various methods to recycle waste materials, including extraction, reprocessing, conversion, and reuse.

In today's culture, recycling refers to the widespread collection of household material and everyday waste items, such as empty paper boxes, cans, and plastic bottles. In architecture, recycling refers to the processing of C&D materials that include: (1) nonhazardous waste, (2) hazardous waste, and (3) mixed material waste that may contain some hazardous components. Although most building material waste contains no harmful components, many states have strict regulations about its transportation, storage, and disposal. The proper management of hazardous waste removal is a contractor's responsibility and is additionally regulated by the U.S. Environmental Protection Agency (EPA).

■ Common C&D waste includes metal, lumber, masonry, glass, plastic, paper, appliances, asphalt, paints, and landscape-related materials. Most of these materials can be reused and/or recycled, which is one of the main tenets of green architecture (see Table 5.3). Green buildings should utilize and reclaim materials from demolished buildings, including salvaging lumber and masonry, using reclaimed aggregates from crushed concrete and drywalls, and reusing appliances, doors, windows, and other fixtures.

PASSIVE SOLAR DESIGN METHODS

Passive solar design involves using nontechnical design methods (e.g., site conditions, local climate, sun angle, building massing and orientation, and daylight) to save and retain energy within buildings. Unlike their technologically advanced active solar counterparts, passive solar buildings do not rely on electrical or mechanical systems, control techniques, or other devices to operate.

There are two basic passive solar design methods: (1) direct gain and (2) indirect gain.

■ *Direct gain.* In the direct gain method, the building is designed to be directly heated by solar thermal energy, and the living space acts as a solar collector, heat absorber, and distributor (see Fig. 5.8). This method relies on the orientation of the building, the location of its openings, the building's materials and their attributes, the structure's heat storage capabilities, and its insulation systems.

In this method, sunlight is allowed to enter the building through south-facing windows. Light is absorbed directly by the thermal mass, which stores and releases the heat as the building cools. Since absorption, retention, and release of thermal energy are the key factors for this method, the physical and chemical components of the thermal mass materials are of utmost importance. For example, earth materials, such as brick walls and stone floors, inherently absorb and retain thermal heat longer than most other building materials. As the building cools at night, the thermal mass radiates the stored heat into the building.

If designed correctly, the direct gain method can utilize up to 75 percent of the solar thermal energy that enters the building. Although this method is quite simple, direct, and effective, it may also produce overheated spaces that can damage furniture and/or create uncomfortable living environments.

TABLE 5.3   COMMON BUILDING MATERIALS AND THEIR RECYCLING VALUE

CONSTRUCTION MATERIAL	RECYCLE	REUSE	INFILL	INCINERATION
Concrete	●		●	
Cinder blocks	●	●	●	
Drywall	●	●	●	
Masonry	●	●	●	
Wood/lumber	●	●	●	
Plaster			●	
Asphalt shingles			●	
Wood shingles	●	●	●	
Steel	●	●	●	
Structural steel	●			
Steel poles	●	●		
Rebar	●	●	●	
Siding	●		●	
Doors/windows	●	●		
Plumbing fixtures	●		●	
Electrical wiring	●			
Appliances	●	●		
Glass	●	●	●	
Sand	●	●		
Soil	●	●		
Landscape	●	●		
Paints/solvents/sealers			●	●
Treated wood			●	●
Asbestos			●	●
Other hazardous materials			●	●

- *Indirect gain*. The indirect gain method requires a buffer thermal mass between the sun and the living space to be heated. The thermal mass buffer can be a structure, a wall system, an absorption device, and/or another space. Unlike the direct gain method, the indirect gain system has the thermal mass act as a collector, absorber, and distributor of the solar energy. Thermal distribution is accomplished via conduction, averaging 40 percent utilization rates. There are three main types of indirect gain systems: thermal storage wall, roof water, and sunrooms.

DAY: ABSORBS SUN

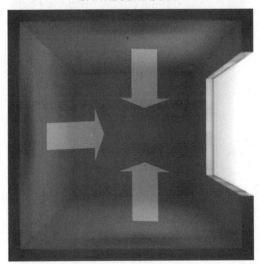

NIGHT: RADIATES HEAT

Figure 5.8   Direct gain is a simple and effective passive solar method, which can utilize up to 75 percent of the solar energy.

■ *Thermal storage wall.* The Trombe wall is the most widely used thermal storage wall method. In this method, a thick masonry wall (10–16 in) is placed on the south side of a building where it will receive the most sunlight (see Fig. 5.9). Dark-colored, single or double glazing windows cover the exterior of the wall to effectively absorb the solar energy. The absorbed energy is stored and then radiated to the living area after the space cools. For other thermal wall systems, the wall thickness varies depending on the material—10–14 in for brick, 12–16 in for concrete, 8–12 for adobe, and at least 6 in for water.

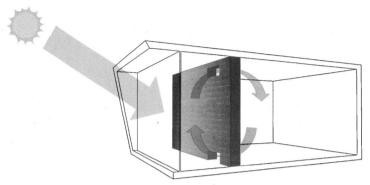

Figure 5.9 Trombe wall is the most widely used thermal storage wall method.

■ *Roof water systems*. Roof water systems absorb and transfer heat from outside to inside during the winter, and from inside to outside during the summer (see Fig. 5.10). In order to heat and cool a building effectively during both seasons, the water stored in tanks or pipes (along with additional treatment materials,

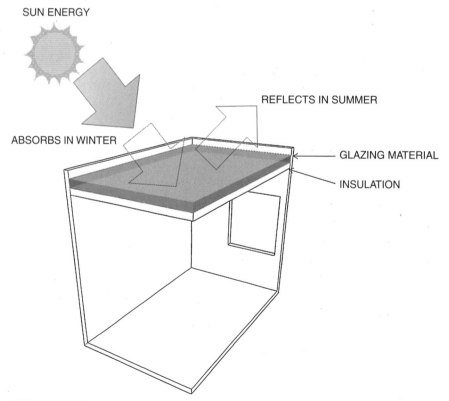

SUN ENERGY

REFLECTS IN SUMMER

ABSORBS IN WINTER

GLAZING MATERIAL

INSULATION

Figure 5.10 Roof water systems are primarily used in commercial buildings in low humidity climates.

such as antifreeze) must be insulated in reverse order from winter to summer. These systems are primarily used in commercial buildings in low humidity climates.

■ *Sunroom.* The use of a sunroom is a hybrid method, which includes the functions and benefits of both the direct and indirect gain methods (see Fig. 5.11). Thermal energy can be gained by any combination of the before-mentioned methods, and then transferred to the living area through conduction from a shared wall, or through wall vents. In the vent system, the amount of heat transferred may be controlled by opening and closing the vents. Also called "solar greenhouses" or "solariums," sunrooms have significant advantages over other indirect gain solutions, especially because of their ability to control the level of heat within a building. As for the direct gain method, however, the excess solar gain within sunrooms can be harmful if not designed correctly.

We can see that the challenge of generating energy and retaining energy in a way that is nonthreatening to our environment is a topic that is as wide-ranging as it is

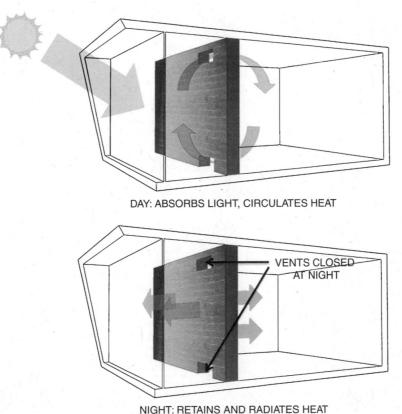

DAY: ABSORBS LIGHT, CIRCULATES HEAT

VENTS CLOSED
AT NIGHT

NIGHT: RETAINS AND RADIATES HEAT

Figure 5.11   **Hybrid method of passive solar gain.**

critical. The time is past for mankind to shrug off these concerns as something that does not affect us. As has been previously stated, architecture is a major contributor to environmental pollution. Introducing architecture students and the general public to the concepts of green architecture can spark awareness, increase the understanding of environmental building methods, and ultimately change our world.

6

GREEN MATERIALS

"Green" materials are so named because they are environmentally friendly, biodegradable, renewable, and recyclable. Green materials can be subdivided into four main categories, the first of which is biomaterials or biotic materials. Included in the biomaterials category we also find natural materials, biopolymers, and bioplastics. The three other main categories are composites, smart materials, and nanomaterials.

Biomaterials/Biotic Materials

Biotic materials are any natural materials (such as bamboo or wood) that originated from living organisms. Biotic materials that have undergone more extensive processing are referred to as bio-based materials. One example of a bio-based material is casein; it is extracted from milk, but then is processed in various ways before becoming an ingredient in products like paint and protective coatings. Therefore, biotic materials can be natural or synthesized organic compounds that exist in nature (Wool and Sun 2005).

NATURAL MATERIALS

Natural materials come from plants, animals, or the earth. They are naturally occurring, renewable, and biodegradable materials, which are excellent for building, as their use requires little construction knowledge. Since builders generally choose locally available materials, their associated embodied energies are relatively low.

Wood Wood is a hard, strong, fibrous tissue composing the stems of woody plants, such as trees and shrubs. This natural material is heterogeneous, hygroscopic, cellular, and anisotropic. It is heterogeneous because the material consists of multiple layers with varied structural properties, mainly fibers of cellulose (40–50 percent) and hemicellulose (15–25 percent), impregnated with lignin (15–30 percent) (Hoadley 2000; Porter 2007). Its hygroscopic nature gives wood the ability to attract water molecules from the surrounding environment through either absorption or adsorption, which makes this material quite moist and water-rich. After being cut and stacked to dry, wood loses water and shrinks, losing from 10 to 50 percent of its volume (Porter 2007). However, because of its hygroscopic nature, wood swells again when exposed to water. Density and moisture content defines the strength, elasticity, and toughness of wood. Because of its anisotropic nature, its properties (such as elasticity, strength, and hardness) vary with differing external stimuli. Since wood is subject to these changes in dimension, stresses can develop with repeated swelling and shrinking (Patrick 1988).

Wood is a traditional natural green material with high levels of sustainability and performance values (see Fig. 6.1). The material's sustainability is derived mostly from its inherent attributes, such as durability, economical nature, resource efficiency, maintenance, and recyclability. Its durability is measured by its structural strength and its resistance to decay and rot. Although the strength of wood depends on its stiffness, density, and more importantly, on the individual species, structurally, wood has high levels of tensile, compressive, shear, and impact values. Wood decay is directly related to the climate and humidity levels. The most severe decay occurs in areas where rainfall is high and weather is warm and humid. A high temperature also affects the strength of wood; for every 1°F increase in temperature, wood correspondingly decreases 1 percent in strength (Hoadley 2000; Porter 2007).

If a particular type of wood is not suitable for construction, it is processed via chemical or mechanical methods and used as raw material for other building products, such

Figure 6.1   Wood accounts for 45 percent of raw construction materials worldwide with significantly low-embodied energy. Wood lumbers are mostly used for framing of buildings.

as chipboard, hardboard, medium-density fiberboard (MDF), engineered wood, and oriented strand board (OSB). Overall, wood is an energy-efficient, strong, durable, recyclable, and flexible material that is naturally sound-absorbent and heat-resistant. Although the measure of its resistance to heat flow, or R-value, changes depending on the species type, wood with open cell structures are the most effective.

Bamboo Bamboo is an organic, natural material that is lightweight, durable, flexible, biodegradable, and recyclable. There are over 1600 species of bamboo throughout the world, 64 percent of which are native to Southeast Asia (Farrelly 1984). Bamboo varies in height from plants measuring only 1 ft to giant hollow stems that can grow to over 100 ft. It is found in various climates, from tropical jungles to cooler high-altitude areas.

Bamboo is the fastest-growing natural material, and most species reach maturity between three to five years. Because of its fast growth rate, bamboo is considered to be a rapidly renewable material resource (see Fig. 6.2). Although bamboo has inherently strong structural properties, it must be treated before being used as a building

Figure 6.2   **Bamboo is the fastest growing renewable material with inherently strong structural properties.**

material. For example, large timber bamboo is used as a strong and durable flooring material. When treated, it becomes as strong as steel and may be used as a main structural material in small dwellings. Because of its availability, strength, and low cost, bamboo is often used as scaffolding material in many Southeastern countries and in Latin America.

Straw bale   Straw bales are an agricultural by-product made from the stems of cereal crops, such as wheat, oats, rye, rice, barley, and others. Its low cost and general availability make straw bales a highly desirable, natural green material (see Fig. 6.3). There are two basic straw bale construction methods—as infill and as load-bearing. A post and beam framework is the most common nonload bearing construction method, where the framework supports the structure and straw bales are used as infill. Load-bearing structural straw bale construction is also prevalent, but special attention and care is required here, as the weight of the roof must be supported by the bales.

Straw bale is a very economical material. Depending on the local availability, straw bale is a resource-efficient, renewable material with very low-embodied energy. Building walls with straw bales do not require skilled labor, and erecting walls can go amazingly quickly. Bales must be protected from getting wet; moisture level should be kept below 14 percent during storage and construction (Lacinski 2000). The bales should also be kept dry at all times after construction.

Straw bale has an excellent insulation value, exhibiting R-values from R-30 to R-50 (Lacinski 2000). The bales are typically covered with concrete mortar/stucco, providing a high degree of fire resistance. When dry, straw bale structures are also

Figure 6.3   Straw bale as an infill green material.

rot-resistant and do not decay. However, extensive care is required to avoid water entering or collecting in straw-bale construction, as this will cause it to rot and decay over time.

Cordwood Cordwood—also called firewood—is a practical, resource-efficient, renewable material that is usually obtained from timber and trees unsuitable for building purposes. In the United States, firewood is sold by the cord (128 ft^3); therefore, it is called cordwood. Cordwood was the primary source of fuel until the 1800s, when it was displaced by coal and later by oil. Some firewood is a by-product of natural forests. Deadfall and standing dead timber are preferred, mainly because they are partially seasoned and contain minimal rot.

Like straw bale, cordwood is mostly used as infill material, although stand-alone cordwood structures have considerable load-bearing strength. As a load-bearing wall, the compressive strength of wood and mortar requires a roofing to be directly connected to the wall (see Fig. 6.4). Cordwood has significant thermal mass capabilities. Because of the central cavity between the inside and outside masonry, mortar is used to cement the logs together. However, different mortar mixtures and construction techniques affect the resistance of heat flow, impact its inherent thermal mass, and change its heat storage capacity.

Overall, cordwood is an inherently green material because of its resource efficiency: using mainly waste material, its high-insulation value, low maintenance, ease of construction with limited or no skills, and very low cost.

Rice hulls Rice hulls are the hard, protective shells formed over rice grains during the growing season. The hull, which is made of 20 percent silica and 30 percent lignin,

Figure 6.4   Cordwood wall construction.

is an agricultural waste product, as it is indigestible to humans. Rice hulls resist moisture penetration, which makes this material a great vapor buffer by absorbing excess moisture and releasing it back to the air as soon as possible. Typically, they hold only about 10 percent of their weight in moisture. Rice hulls are excellent insulation materials, with a thermal resistance of about R 3.0 per inch. They are fire resistant, and also resistant to rot, molds, mildew, insects, rodents, and fungus.

Overall, rice hulls are an excellent green material because of their resource efficiency; it is a practical use of waste material that has a high-insulation value, yet is low maintenance. An unskilled worker finds the material easy to use, and there is a very low cost. Moreover, rice hull construction exhibits excellent load bearing capability, sound absorption, fire resistance, and (when made into hollow interlocking panels) is suitable as a building material in areas prone to earthquakes.

Mud and clay   Aside from tents made of cloth or leather, the most basic building materials are mud, rocks, and clay. People all over the world still use these natural materials to create shelters to suit their local weather conditions. Regardless of the choice of other basic materials as structural components (e.g., stone and brush), mud is typically used as a type of concrete and insulation. Depending on the building style, the mixture and the amount of these materials vary. The cob/adobe style requires greater amounts of clay, whereas low clay soil is usually associated with sod building.

Mud and clay have excellent thermal mass qualities and keep the indoor temperatures at a constant level. Buildings made of mud and clay are naturally cool in the summer heat and warm in cold weather. In this ability to hold the gained temperature over a long period of time and release it slowly, it can be compared to stone. Overall, mud and clay are highly available, low-cost, energy-efficient, high-performance, ecologically green materials, but they may require constant maintenance and structural support.

Sod   Sod is a thin block of grass held by its roots, usually used for turf and lawn establishment. It can also be used as a temporary building material, especially in grassland areas where other materials are unavailable. Like brick, sod is cut and laid in regular block shapes, but their shapes are different than bricks, being wider and stronger. Sod was a common construction material, especially during the nineteenth century in the settlement of the North American prairie (see Fig. 6.5).

Because of its organic composition, sod can easily be damaged by external factors and natural conditions such as rain, snow, strong winds, and erosion. Therefore, the outer walls of a sod building are usually protected with a layer of stucco or wood panels. Similarly, bare sod inside the walls presents dirt, smell, and is open to bacterial contamination. Therefore, the interior is typically covered with a protective material, such as tar paper or plaster.

The advantages of sod are its availability, affordability, and insulation value, which make sod houses cool in the summer and warm in the winter. It is also fireproof. The main disadvantages are that sod is heavy, hard to cut, tends to be damp, and deteriorates quickly, thereby requiring constant maintenance.

Figure 6.5 Sod house in the prairie, circa 1900,
Saskatchewan, Canada.

Rammed earth Rammed earth is a compressed damp mixture of soil mixed with
sand, gravel, clay, and other stabilizers. It is pressed against an external support frame
and creates a solid earth wall by molding the shape of a wall section. Although the
main constituent is sand, other material amounts in the mix (i.e., gravel, clay) may
vary. The ideal proportion of rammed earth is at least 50 percent sand and 15 percent
clay; lower proportions may cause a wall to collapse, while higher proportions cause
shrinkage (McHenry 1989).

Rammed earth is an economical and versatile material, which is as effective for
nonlinear surfaces, corners, curves, and arches, as it is for straight walls. Although the
material costs are low, rammed earth construction requires skilled labor, which makes
buildings composed of this material comparable in cost to conventional constructions.

Overall, rammed earth is a clean, recyclable, biodegradable, resource-efficient
material with a high-thermal mass value. Since the material uses the subsoil as the
main ingredient, it can absorb and store solar energy during the day and radiate it
slowly into the building at night. Its inherent insulation and radiation capability
reduces the energy required for heating, and further reduces carbon dioxide emissions
and pollution. Rammed earth has low-embodied energy, as the earth can be dug
locally, thereby reducing transportation and manufacturing requirements.

BIOPOLYMERS

Biopolymers are a class of polymer molecules derived from living organisms.
Polymers are high molecular–weight substances composed of repeating subunits
called monomers (Kaplan 1998; Wool and Sun 2005). The structure of biopolymers
differs, based on the chemical composition and sequencing of its monomers, but it is
inherently well defined. Its biological functions are determined by how the biopolymers spontaneously create compact shapes.

Biopolymers occur naturally, being produced by plants, animals, and microorganisms through biochemical reactions. A wide range of naturally occurring polymers from renewable resources is available for material applications (Kaplan 1998; Yu 2008). Biopolymers are produced without toxic waste and are inherently biodegradable; therefore, they are considered to be a green alternative to petroleum-based polymers that take a long time to biodegrade. Most biopolymers are produced by extraction from the natural plant or animal material, followed by a purification process. Some biopolymers can be made commercially, using microorganisms like bacteria and fungi, through a fermentation process. Microbial production of biopolymers on a very large commercial scale has been developed relatively recently (Kaplan 1998; Wool and Sun 2005; Yu 2008).

Some biopolymers that are not found in nature can still be produced commercially from monomers that are found abundantly in nature. The resulting polymers have the properties of naturally produced biopolymers, and are referred to as "honorary biopolymers" (Stevens 2002). Polysaccharides, proteins, and polyesters comprise a large and chemically diverse supply of renewable polymers, ready for use in the production of biodegradable plastics (Stevens 2002).

Biopolymers are used extensively in architecture and the building industry in widespread, diverse applications. Although some biopolymers can be applied directly, most of them are used in indirect applications or to optimize other material properties. For example, concrete and dry-mix mortars (such as adhesives and plasters) represent two major types of applications for biopolymer use in buildings. Other major uses of biopolymers are cellulose and lignin as insulation, natural animal fibers and plant seeds as enclosures, and polymerics in pipes and pipe fittings.

Cellulose   The most abundant type of biopolymers, which make up approximately 75 percent of all organic matter are called polysaccharides (Bersch 2009). Cellulose, the structural component of the primary cell wall of green plants, is the most common of these polysaccharides.

Cellulose is a resource-efficient, clean, recyclable, biodegradable material with a low-embodied energy, and high-insulation value. It requires 20 to 40 times less energy to produce cellulose insulation materials than it does to produce furnace-based insulation materials (Bersch 2009). Recycled cellulose insulation material is approximately 80 percent postconsumer recycled newspaper by weight; the rest is comprised of fire-resistant chemicals and—in some products—acrylic binders.

Cellulose insulation has a very low pollution rate, as it does not contain formaldehyde-based glues, which present a continuing toxicity after installation due to off-gassing of formaldehyde. Although it may initially cost more than fiberglass insulation, cellulose insulation is very economical, as it can save homeowners up to 50 percent on their utility bills if properly installed. Moreover, cellulose is fire resistant, fills irregular and hard-to-reach places, and is a green choice for consumers interested in environmentally safe products.

Starch   Starch is an abundant component of the planet's biomass and is found in corn, potatoes, wheat, and other plants. The major polymer components of starch are amylopectin (72 percent) and amylase (28 percent). The food sector uses over half of produced starch to manufacture corn syrup and other sweeteners, or as feedstock (raw material) for various manufacturing processes in the chemical, pharmaceutical, and brewing industries. Most of the remaining production of starch is used for nonfood industrial purposes in the manufacture of paper and cardboard, paper coatings, textile and carpet sizing, and adhesives (Stevens 2002).

As a building material, starch is an important component of plasterboard, mineral fiberboard, and adhesives like wallpaper paste and glues for chipboard. For plasterboard or gypsum wallboard, starch is used as a core binder to protect the gypsum crystals from dehydration and breakage during periods of high temperatures. Starch is gelatinized during manufacturing to provide a strong bond between the board and the liner.

Lignin   Lignin, an integral part of the cell wall of plants, is extracted from renewable natural resources like trees, plants, and agricultural crops. After cellulose, lignin is the second most abundant component of woody material; it accounts for about 20 percent of all organic matter (Stevens 2002). Lignin is a clean, nontoxic, versatile product used in the industrial and food processing industries. Like starch, lignin has various industrial uses such as: binders (in ceramics, dust suppressants, plywood and particle board, fiberglass insulation, and soil stabilizers), dispersants (in cement mixes, clay and ceramics, dyes and pigments, concrete admixtures, and gypsum board), emulsifiers (in asphalt emulsions, pigments, and dyes), and sequestrants (in cleaning compounds and water treatments for boilers and cooling systems).

Polyester   Naturally occurring polyesters, like PHA (polyhydroxyalkanoate), are produced by bacterial microorganisms through fermentation. They are biocompatible as well as biodegradable, and have biomedical applications in areas such as controlled-release drugs and surgical sutures. These natural polyesters are used for biodegradable plastics that are suitable for adhesives, coatings, molded goods, and highly elastic products.

Protein   Proteins are formed when amino acids are condensed, resulting in polymerization. There are approximately 20 amino acids that are common to all living organisms, some of which (like collagen and keratin) have structural functions, while others are enzymes. Proteins are used as renewable, biodegradable, green raw materials for an increasing number of industrial applications. These uses are briefly summarized below (Stevens 2002).

Gelatin is used as a thickener for desserts, ice cream, confectioneries, and baked goods. As a hot melt adhesive it has excellent binding properties and is widely used in bookmaking. It has other miscellaneous industrial applications, including its use as a stabilizer-binder in photographic emulsions.

Casein is found in milk and is used in the adhesive on labels in the bottling industries, due to its excellent rheological properties (the way the substance flows). It is also important in binders, protective coatings, and leather finishes.

Whey is the soluble liquid remaining after milk has been separated from the casein curd during cheese manufacturing. Whey protein, derived from whey, has several important applications in the food industry, including acting as an agent that adds body and bulk.

Plant proteins, including soy protein, zein from corn, wheat gluten, potato proteins, and pea proteins, are also important commercially. Soybeans consist of approximately 30 to 45 percent protein and 20 percent oil. Soy protein is used for making plywood adhesive and coatings for paper and paperboard. Corn zein is used as a binder in printing inks, as a shellac substitute, for floor coatings, and in coatings for grease-proof paper. Zein can also be formed into fibers and films that are tough, glossy, and scuff resistant. It is water insoluble and thermoplastic.

Natural fibers   Natural fibers are a type of biopolymer which covers a broad range of vegetable, animal, and mineral fibers. These fibers contribute to and enhance the structural performance of their hosts, and, when used in plastic composites, can provide significant reinforcement.

Natural fibers are mostly used in various composites and play an important role in the building industry. Because of their nontoxic and biodegradable nature, they are green alternatives to timber, concrete, steel, and glass fibers. Various types of natural fiber composites (e.g., rice and coconut husk, groundnut shell, and cotton stalk) may be used as green building materials for numerous applications. Rice husks may be used as an energy source for the production of acid proof cement and rice husk binders. Because of its durability, coconut husk can be manufactured into insulation boards and roofing sheets. Groundnut shell is used to make chipboards and particleboards, and cotton stalk may be used to plaster walls.

Natural fibers are durable, lightweight, and corrosion-resistant materials with a high strength and stiffness, and very low-embodied energy. These biopolymers are highly economical, recyclable, nontoxic, and efficient.

BIOPLASTICS

Bioplastics are a form of biodegradable plastic derived from renewable plants, such as soybean oil and corn starch, as opposed to fossil fuel plastics, which are derived from petroleum. Biopolymers are the base materials for bioplastics, which must contain at least one biopolymeric substance as the main ingredient, and supported by plasticizers and additional additives as necessary (Stevens 2002). The most common types of bioplastics are cellulose-based (CBP), starch-based (SBP), polylactic acid (PLA), poly-3-hydroxybutyrate (PHB), and polyamide 11 (PA11).

One of the key advantages of bioplastics is that their production requires practically no fossil fuels. Their disposal releases about the same amount of toxic gases as is absorbed by the plants from which they originate. Another advantage is availability of raw material resources, such as sugar and starch, for their production. Due to the similarity to conventional plastics and to the environmental benefits, bioplastics are rapidly gaining popularity in architecture. Currently, conventional plastics are one of the most widely used materials in the building sector. Globally, 30 percent of all plastics are used for products in the building industry, ranging from piping to floor covering, which places the building industry as the second highest user after the packaging industry. The availability of bioplastics is particularly important in architecture because about 1/4 of construction debris ends up in landfills. Moreover, the bulk of electrical, water, and sewage infrastructures currently use toxic thermoplastics as the primary material within the buildings.

Types of bioplastics

1 *Cellulose-based bioplastics.* Cellulose-based bioplastics (CBP) plastics are usually produced from wood pulp. They are used to make film packaging products, such as wrappers and to seal in freshness for ready-made meals.
2 *Starch-based bioplastics.* Because of their inherent ability to absorb humidity, starch-based bioplastics (SBP) are primarily used in the field of pharmaceuticals. Thermoplastic starch is one of the most widely used bioplastics.
3 *Plastarch material.* Plastarch material (PSM) is a new generation bioplastic material. It is a biodegradable thermoplastic resin composed of modified cornstarch combined with other biodegradable materials. PSM has a low production cost due to the availability and quantity of its raw material; it has an excellent degradation performance; and it is capable of withstanding high temperatures. It also has good resistance to water and oil, and is renewable. PSM is a biodegradable, stable material with reasonable softening and melting temperatures (257°F and 313°F). PSM is currently used for a wide variety of applications, such as temporary construction tubing, construction stakes, window insulation, industrial and agricultural film and foam packaging, and plastic bags.
4 *Polylactic acid.* Polylactic acid (PLA) is a highly versatile bioplastic material, which is derived from various bioresources such as potato, corn starch, or other starch-rich substances like sugar or wheat. PLA has similar characteristics to common plastics such as polyethylene, polypropylene, and PET, and can be processed using the same production equipment. Its clarity and strength make it useful for biodegradable products like packaging, lawn-waste bags, coating for cardboard, and fibers for carpets, wall coverings, sheets, and towels. It is also used for biomedical applications, such as sutures and prosthetic materials.

Some of the drawbacks of PLA are the need for fossil fuel in its production, the extended length of time required for degradation within waste streams, and cost. Although the completed product is safe and nontoxic, PLA releases carbon dioxide and methane during the biological breakdown phase. PLA is biodegradable to a certain extent, but it requires ideal conditions to decompose. For example, the decomposition rate is much slower in commercial and municipal facilities than under ideal laboratory conditions, where the process can be controlled. Worst of all, PLA does not degrade in landfills because of the lack of sufficient moisture and correct temperature. Cost is another problem with PLA. Although its price varies with the amount being produced, PLA is still more expensive than traditional plastics.

5 *Polyhydroxy-alkanoates.* Polyhydroxy-alkanoates (PHA) is a polyester bioplastic, which is produced in nature by bacterial fermentation of sugar or lipids. The production of PHA requires no additional equipment cost, as it is able to be processed on conventional processing machinery. It is strong, UV stable, and has a good resistance to moisture and odors.

PHA does have some drawbacks, however. One problem is its lack of flexibility. Depending on its composition, this material is stiff and brittle. Another drawback is the production cost of the fermentation and purification processes that are required. Recent research into lowering such production costs is aimed at genetically modifying plants so as to produce the polymer by farming methods as opposed to fermentation processes. In 2005, Metabolix, Inc., received the United States Presidential Green Chemistry Challenge Award in the small business category for their development and commercialization of a cost-effective method for manufacturing PHA.

Another possibility is PHA produced by microorganisms, which would have potential applications within the medical and pharmaceutical industries, primarily due to its biodegradability.

6 *Poly-3-hydroxybutyrate.* Poly-3-hydroxybutyrate (PHB) is another biodegradable polyester produced from plants and other renewable raw materials such as glucose, agricultural waste, and sugar beets. Although PHB is very similar to the solid plastic polypropylene used in products ranging from food packaging to textiles, it has advantageous differences. PHB is transparent, nontoxic, and biodegradable without residue. Other advantages are its temperature resistance and ease of manufacturing. Like PHA, PHB can also be processed on conventional processing equipment with no additional cost.

While PHB is 5 times more expensive than traditional plastics, it is water insoluble and relatively resistant to hydrolytic degradation. This differentiates PHB from most other available biodegradable plastics, which are either water soluble or moisture sensitive.

7 *Polyamide 11.* Polyamide 11 (PA11) is a high-performance bioplastic, which is, derived from vegetable oil and is known by the trade name Rilsan(r) PA. It has a unique combination of properties, such as: outstanding chemical resistance, excellent durability, customizable levels of flexibility, superior cold impact resistance, excellent gas and liquid barrier, and good abrasion resistance. These features make PA11 valued for use in car fuel lines, pneumatic air brake tubing, electrical antitermite cable sheathing, oil and gas flexible pipes, and control fluid umbilical. These are often reinforced with fibers from the kenaf plant to increase heat resistance and durability.

PA 11 is lighter and more resistant to stress-cracking than traditional plastics, making it ideal for high-performance window and other transparent enclosure applications. It has high UV resistance, a better dimensional stability since it absorbs less moisture, and may be easily mixed with high-performance elastomeric products.

As has been seen, there are a great many types of biomaterials/biotic materials that are considered "green." We will now turn our attention to composite materials, the second of the four categories into which green materials are divided.

Composites

Composites are engineered materials formed by combining two or more different elements with dissimilar properties. These different combined elements, when observed through a microscope, are seen to have remained separate and distinct within the finished structure, yet the combination creates a completely new material. The properties of the newly designed composite are different from those of the original constituent materials acting independently; in fact, the new composite appears to have taken advantage of the different strengths and abilities of the elements from which it was formed (Daniel and Ishai 2006). Composites consist of two elements: (1) matrix (or binder) and (2) reinforcement.

The matrix is the body constituent of a composite that completely surrounds the dispersed element and gives the body its bulk form. A composite may have a ceramic, metallic, or polymeric matrix that is present in greater quantities within the composite most of the time (Matthews and Rawlings 1994). The reinforcements are harder and stronger than the matrix and enhance the properties of the matrix. The resulting composite consists of layers of reinforcements and matrix stacked in such a way as to achieve the desired properties (Schwartz 1997). For example, in the case of mud bricks, the two roles are taken by the mud (i.e., matrix) and the straw (i.e., reinforcement); in concrete, by the cement and the aggregate; in a piece of wood, by the cellulose and the lignin. In fiberglass, the reinforcement is provided by fine threads or fibers of glass, often woven into a sort of cloth, and the matrix is a plastic.

Green composites are similar to regular composites, but they are designed with the lowest environmental footprint possible (Baillie 2004). The raw materials of green composites are not necessarily derived from renewable resources, mostly because of financial reasons. For example, most synthetic polymers are currently derived from petrochemical resources. Some natural materials require considerable purification and processing before they can become suitable for use in composites. Therefore, many raw materials are semisynthetic: a combination of natural and renewable elements (Baillie 2004). Green composites are inherently biodegradable, but their biodegradability is dependent on conditions of time, temperature, and local conditions. The biodegradability of the composite is only desirable if it does not harm the properties of the composite while in use (Baillie 2004; Shanks 2004).

In architecture, designs using composite materials have been the principal manner of making functional products for ages. Though restricted to natural materials, tremendous sophistication and technical expertise characterize many indigenous and preindustrial objects and methods. These skills have been developed over many generations as humans invented, refined, and perfected designs into what could be termed a "stable technology" (Rose 2004).

COMPOSITE CLASSES

Biocomposites   Biocomposites are inherently renewable, recyclable, and biodegradable. In comparison to other composites, they are economical, nontoxic, and require much less embodied energy to produce. They have several advantages over traditional composites, including low cost, low density, acceptable specific strength and stiffness, enhanced energy recovery, carbon dioxide sequestration, and the need for much less embodied energy to produce. Their environmental impact is significant beyond the toxicity and the landfill. For example, the high-fiber plants commonly used for composites are easily grown, require few pesticides, and can be rotated with traditional food crops. These composites are used as adhesives, films, foams, rigid and flexible panels, coating, resins, and elastomers. In architecture, biocomposites are widely used for architectural applications in making building products (such as decking, roofing, doors, and windows) and in structural and nonstructural assemblies (Femandez 2002).

Ceramic composites   Ceramic materials have high temperature resistance but also have limitations in structural applications because of their brittle composition. The incorporation of ceramic-fiber reinforcement into the ceramic matrix improves this structural limitation. The enforcements can be continuous or chopped fibers, small discontinuous whisker platelets, or particulates. They can be used for high temperature applications. The average fracture toughness, which is the major weakness of ceramics, is doubled with ceramic-matrix composites (Barsoum 1997; Schwartz 1997; Chawla 2003).

These composites are mostly used in the aerospace industry. The major drawback, however, is that the high-temperature reinforcements oxidize in the air.

Advantages
- High strength and strength-to-density ratio
- High strength at high temperature
- Low density
- High stiffness-to-density ratio
- Toughness (impact to thermal shock)
- Improved fatigue strength
- Improved stress rapture life
- Controlled thermal expansion and conductivity
- Improved hardness and erosion resistance
- Ability to be tailor-made with customized properties
- Ability to be fabricated into complex components to near net shape

Disadvantages
- Brittleness
- Cost
- Lack of structural, load-bearing abilities

Application areas
- Military technologies
- Aerospace industries

- Auto industries
- Sports
- Electronics
- Building industry

Polymer composites Polymer-matrix composites are the most advanced and developed class of composites with a wide range of applications, but especially where large and complex forms are required. They consist of high-strength fibers (such as glass and carbon) in a thermoplastic resin. Polymer composites are strong, durable, flexible, high-strength composites with a very high resistance to weathering and corrosion. One of the main advantages of polymer composites is their ease of shaping during manufacturing. They easily adopt flat, curved, or sharply sculpted contours, thereby showing tremendous design flexibility.

Currently, these composites are used in a broad range of applications. In the aerospace and automotive sectors, they are mainly used for fuel efficiency because of their lightweight. Because of the curvilinear shapes of the products in these sectors, manufacturers are also taking advantage of its design flexibility, and part consolidation that these polymer composites can provide. In the energy sector, the growing demand of wind energy has led to increased interest for polymer composite turbine blades. In the building industry, their durability, flexibility, ease of shape, and manufacturing make them stiff, lightweight alternatives to traditional building materials, such as steel, aluminum, and wood. Moreover, their resistance to corrosion enables their use in marine, construction, and infrastructure applications, including piping and storage tanks.

Advantages
- High tensile strength
- High stiffness
- High fracture toughness
- Good abrasion resistance
- Good puncture resistance
- Good corrosion resistance
- Good chemical resistance
- Low cost
- Low maintenance
- Construction simplicity
- Versatility
- Load-bearing capacity

Disadvantages
- Low thermal resistance
- High coefficient of thermal expansion
- Not stiff in the perpendicular direction

Application areas
- Building/construction
- Medicine
- Aerospace
- Automotive
- Civil/marine engineering

Metal composites   Metal composites consist of metal alloys reinforced with continuous fibers. These composites have a high stiffness, durability, and temperature resistance but in general they are heavier than many other composites. Although their use is not as common as the polymer composites, metal composites are finding increasing application in many areas.

In architecture, metal composites provide a low-maintenance, sturdy, and durable structural building material that can significantly reduce the time and complexity of construction. These composites can be formed into various shapes and curves by joining together in a limitless combination of geometric combinations. Metal composite materials were originally called aluminum composite materials, which is still the predominant material. The name was later changed to metal composite materials to reflect the introduction of new skin materials, such as copper, stainless steel, and titanium.

The advantages of metal composite materials are their high-tensile strength and shear modulus, high-melting point, low expansion, resistance to moisture, dimensional stability, ease of joining, high ductility, and toughness (Schwartz 1997). They are among the most consistent and precise construction products available. For example, metal composite sheets remain flat during manufacturing and after installation. Since the skins are bonded to a core plastic under tension, a balanced panel is produced without deformation or wrinkling.

Advantages
- High strength and toughness
- High stiffness
- High shear strength
- Wear resistance
- Abrasion resistance
- Fire resistance
- Low coefficient of thermal expansion
- Good electrical conductivity
- Low density
- Durability
- Low maintenance
- Shape flexibility
- High ductility

Disadvantages
■ Poor resistance to seizure and galling
■ Higher cost of some material systems
■ Relatively immature technology
■ Limited service experience

Application areas
■ Building/construction
■ Aerospace
■ Automotive

Hybrid composites   Hybrid composite materials represent the newest of the various composite materials currently under development. The hybrid composite category covers both the hybridizing of a composite material with other materials (either other composites or base unreinforced materials) and composites using multiple reinforcements. Furthermore, this category covers the use of multiple materials (at least one of which is a composite) in structural applications and highlights the multiple uses and advantages of composite materials (Schwartz 1997).

Carbon fiber composites   Carbon fiber is an incredibly strong, lightweight material that is primarily composed of carbon atoms and ultrathin fibers (see Fig. 6.6). It is often used to reinforce composite materials, particularly the class of materials known as carbon fiber or graphite reinforced polymers. Nonpolymer materials can also be used as the matrix for carbon fibers. Due to water solubility and corrosion considerations, carbon has seen limited success in metal matrix composite applications. The alignment and structure of the fibers and the low density of the composite provide

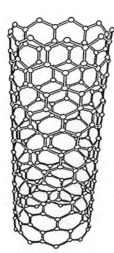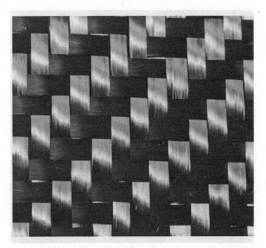

Figure 6.6   **Twill weave carbon-fiber composites are the most commonly used weave patterns in the industry today.**

extra strength and lightness to the material. Therefore, carbon fiber is much stronger and lighter than fiberglass and aluminum, and twice as stiff, 10 times stronger, and lighter than steel.

Because of these strong and highly engineered properties, it is mainly used within the automotive, aerospace, and sports equipment fields, where strength-to-weight ratio is critical. More recently, carbon fiber has also gained acceptance in the construction industry as a replacement for traditional structural reinforcement, with applications ranging from seismic strengthening of concrete columns and bridge girders, to foundation reinforcement.

The disadvantages of carbon fiber are its cost and its lack of reusability. Carbon fiber is 10 times more expensive than steel and other compatible metals. Reusability and recyclability are additional issues, creating a waste disposal problem. Carbon fiber cannot be reused and it loses strength when recycled.

Ethylene tetrafluoroethylene   Ethylene tetrafluoroethylene (ETFE) is a fluorocarbon-based polymer that possesses a unique combination of properties. It has outstanding physical strength and durability yet it is lightweight. It is nontoxic, and is resistant to corrosion, chemicals, and heat.

ETFE has an excellent strength-to-weight ratio and elasticity. It can bear 400 times its own weight, can be stretched to 3 times its length without loss of elasticity, and yet weighs only 1 percent of a comparable glass panel. Its nonstick surface is corrosion- and dirt-resistant, and it fights chemical exposure making ETFE a self-cleaning material. ETFE also has high resistance to heat, low smoke and flame characteristics, and significant resistance to weather and aging. Another advantage of this material is its cost-effectiveness; installation costs for ETFE are between 1/4 and 3/4 times less than for regular glass panels.

The main disadvantages of this composite are its transitivity of light and sound, maintenance, and fabrication complexity. Although it is more advantageous than glass in many aspects, ETFE transmits more light and sound than glass. This material also requires constant maintenance. ETFE is not tear-resistant; therefore, it is mostly used as a higher level enclosure and roofing material. If the material is torn, however, it can be patched easily with other pieces of ETFE. In most building applications it is used as ETFE bubbles, where multiple layers are pressurized to form a composite unit, sealed, and held by metal composite strips such as aluminum.

High-density polyethylene   High-density polyethylene (HDPE) is a hybrid polymer composite with outstanding properties, such as high strength, high stress, and chemical resistance. This composite is flexible, impact and wear resistant, and durable. HDPE is widely used in wood plastic composites, binding mortars, and in pyrotechnics. HDPE mortars are preferred to steel and plastic tubing because they are more durable, less toxic, and, therefore, much safer than PVC.

HDPE is used in ventilation ducts within commercial and industrial facilities because of its chemical resistance, reliability, durability, and environmental safety. HDPE's pressure loss in ventilation piping can be reduced as much as 25 percent compared with other plastic and metal piping. Due to its light weight, HDPE building products are easy and less expensive to transport, install, and support.

The disadvantages of HDPE are its poor weather resistance and dimensional accuracy, low mechanical and thermal properties, high thermal expansion, and its sensitivity to thick sections. When exposed to ultraviolet (UV) radiation from sunlight, HDPE can suffer surface damage. HDPE is also highly permeable; therefore, it should not be laid in contaminated areas.

Green concrete   Although concrete is a very versatile building material, it has serious environmental hazards. For example, the production process for Portland cement requires heating limestone up to 1400°C, producing double the amount of carbon dioxide (CO_2) per ton of cement when compared to green alternatives. Additional CO_2 and other gasses are produced by its excavation, transportation, and application.

Green concrete is an environmentally friendly version of concrete produced with either non-CO_2-emitting materials and/or non-CO_2-producing methods. There are various different ways to make green concrete such as: producing from abundant, natural resources; using recycled or reclaimed industrial materials; changing the chemical structure to prevent CO_2 emission during combustion; or absorbing CO_2 from the air after application.

One of the most commonly used methods to make green concrete is to use industrial waste by-products such as fly ash (from coal combustion) and blast furnace slag (from iron manufacturing) to constitute the cement mixture used in producing concrete. One of the main advantages of this method is that it prevents these waste materials from entering landfills. Other advantages are the improvement of the strength and durability of the concrete, and the reduction of the CO_2 embodied in concrete by as much as 70 percent.

Green PVC alternatives   Because of its versatility, cost, ease of use, and applicability, polyvinyl chloride (PVC) is one of the most commonly used synthetic plastics in the building industry. However, PVC is a highly toxic material with unsafe by-products, which pose serious environmental and health hazards throughout its life cycle. Moreover, PVC is a recycling contaminant; it is very difficult to recycle and it interferes with the recycling of other plastics.

In the building industry, PVC is used everywhere, but mainly in piping, siding, electrical insulation, sheathing, roofing, door and window frames, wall coverings, flooring, and other miscellaneous uses (carpet fibers and mini blinds). Green alternatives to PVC materials are the following:

- *PVC piping*. Clay, cast iron, and high-density polyethylene (HDPE)
- *Vinyl siding*. Wood, acrylic, and fiber-cement boards
- *Electrical insulation*. Low-density polyethylene (LDPE)
- *Sheathing*. Halogen-free, low-smoke polyethylene and low-density polyethylene (LLDPE).
- *Roofing*. Soil and grass, light metal, thermoplastic polyolefins, and ethylene propylene diene monomer (EPDM)

- *Flooring.* Natural biomaterials such as wood and bamboo, and ceramic composite tiles
- *Wall coverings.* Biofiber and polyethylene
- *Window and door frames.* Wood, fiberglass, and aluminum
- *Carpet.* Recycled biomaterials and natural fiber backing

Green surface composites

Cork composites   Cork surface composites are durable, low-maintenance, decay-resistant materials with excellent thermal and acoustical insulation properties. This composite is made of cork and recycled rubber, and by nature is very soft and comfortable. Its cellular structure traps air inside, which reduces noise and vibration, and gives floors natural shock absorption.

These composites are inherently healthy materials with properties that are antiallergenic and resistant to insects. A naturally occurring substance in cork called "suberin" repels insects and molds, and protects cork from rotting when moist. In addition, these composites are naturally fire resistant and burn with no toxic off-gassing.

Linoleum   Linoleum is a green surface composite made from organic, biodegradable materials—mainly wood flour, linseed oil, and pine resin. The composite structure consists of linseed oil as the binder, lime as the filler, and organic pigments as the color for the material. Natural fibers such as jute are used for stabilizing the material.

Linoleum is a very comfortable, durable, long-lasting material that is easy to maintain. It is also energy efficient. Organic linoleum has significantly low-embodied energy, creates little waste during manufacturing, and can be fully recycled and reused. Since linoleum does not require constant maintenance (such as sealing, waxing, or polishing like other similar materials like vinyl), it is also quite cost-effective. Health is another advantage of linoleum, which has very low VOC emissions when installed with low-VOC adhesives, and it does not contain formaldehyde, asbestos, or plasticizers.

Recycled rubber composites   Recycled rubber composites are widely used green materials that are as strong, durable, and resilient as original tire rubber. It is an ideal green composite for indoor and outdoor uses, and recently has been used for paving materials, such as sidewalks and vehicular roads. Indoor rubber composites are quite sustainable, slip resistant, and require very low maintenance. Its low VOC emissions create healthy environments for the occupants, and it meets U.S. indoor air quality specifications. The surfaces can also be sealed and finished to ease cleaning and maintenance.

These composites are used for all types of applications where falls and injuries may occur, such as playground surfacing, skateboard and swim parks, sport courts, stairways, and rock climbing rocks. Additionally, recycled rubber composites are also used for wall panels, insulation, carpet underlayment, and roofing. The use of this composite as a roofing material has significant green advantages. Because of its chemical structure, it can be molded into different shapes to create various shingle styles, from cedar to slate. It is also water and rust resistant, and withstands extreme temperatures that cause curling, splitting, cracking, and rotting (negative attributes associated with existing roofing materials). One of the main contributions of this material to the environment is that it reuses old tires, which would otherwise end up in landfills and/or contaminate the environment. In addition, rubber composite materials are lightweight, impact resistant, and energy efficient with high-insulation values.

Recycled glass/ceramic Recycled glass/ceramic composites comprise 60 percent of recycled glass products, such as light bulbs, glass waste, and automobile windshields. Crushed glass is mixed with ceramic materials to produce the composite, creating a material that is cost-effective, durable, scratch and wear resistant, easy to clean, and low maintenance.

Recycled wood/plastic composites Recycled wood/plastic composites comprise mainly wood fibers and waste plastics that include high-density polyethylene and other recycled materials. The material is formed into both solid and hollow profiles, and used to produce building products such as decking, door and window frames, and exterior moldings. They are more durable and less toxic than preservative-treated alternatives.

Since wood fibers are used as reinforcement, these composites are heavier and more rigid than regular recycled plastic lumber. The plastic encapsulation and binding makes this composite moisture and rot resistant, but more flexible in hot weather and more rigid in cold weather than other traditional materials.

Hybrid frame systems Hybrid frame systems are a new generation of wood/polymer composites that are extruded into a series of shapes for window frames and sash members. These composites can be treated like regular wood. They are very stable and have the same (or better) structural and thermal properties as conventional wood, but with better moisture and decay resistance.

Hybrid composite panels These composite panels are made from an aluminum core, coupled with wood, polymers, woven fiberglass, and fibers. They are very rigid, lightweight, high-strength materials with excellent impact resistance. These composites are safe, easy to install, and have lower maintenance and costs when compared to traditional paneling structures.

Autoclaved aerated concrete blocks Autoclaved aerated concrete (AAC) blocks are stronger, lighter, green alternatives to conventional masonry blocks. They can be produced from fly ash, which make this composite ecofriendly and nontoxic. AAC is also fire- and insect-resistant, and has excellent sound absorbing properties and insulation values. However, this material has structural limitations—while AACs may be used successfully for low-rise buildings, in high-rise buildings they are primarily used as curtain walls and/or partitions.

Green form-release agents   Green concrete form-release biocomposites are non-petroleum alternatives to conventional oil products. They are typically water-based, nontoxic, biodegradable composites that contain very low VOC. These composites can be used for various types of concrete forms and liners (e.g., steel, aluminum, fiberglass, plywood), and they are useable at below freezing points.

7

SMART MATERIALS

Smart materials are engineered materials that sense and react to environmental conditions, and/or have one or more properties that can be significantly altered in a controlled fashion by external stimuli. These stimuli may include light, temperature, moisture, mechanical force, and/or electric or magnetic fields (Addington and Schodek 2005). All changes are reversible, as the materials return to their original states once the external stimuli expires. The reactions of smart materials vary based on their inherent properties, molecular alterations, and embedded control systems. For example, some materials perform phase changes (i.e., thermochromic) and some perform energy exchanges (i.e., photovoltaic).

Accordingly, there are three types of smart materials: (1) thermo-responsive materials; (2) light-responsive materials; and (3) stimulus (force)-responsive materials. Briefly, thermo-responsive materials undergo transformations due to the changes in temperature. Thermochromic, thermotropic, thermoelectric, and shape memory are primary examples of thermo-responsive smart materials. Light-responsive materials respond to changes in UV light. Major smart materials in this category are photochromic, photovoltaic, and photoluminescent materials. Stimulus (force)-responsive materials are the least defined and most complex among all types of smart materials. These materials undergo transformations due to external stimulants such as electricity, mechanical force, magnetic force, and kinetic energy.

Thermo-Responsive Materials

Thermo-responsive materials (TRMs) are smart materials that transform due to a change in temperature.

THERMOCHROMIC

Thermochromic materials change color in response to temperature differences. When a thermochromic material absorbs heat, its molecular structure and the consequent light reflection of the material also changes. This results in reversible color reflections.

Although there are various different thermochromic compounds, there are primarily two general types: thermochromic liquid crystals and leucodyes. The necessity for temperature precision dictates when each material should be used. When accurate temperatures are needed, thermochromic liquid crystals are used. Liquid crystal molecules are structured on separate layers so that they can change their orientation if a temperature reading differs. A temperature change triggers the molecular change on the crystal, which affects its color reflection from light.

Leucodyes paints or inks work on the same principle and are used in circumstances where precise temperature readings are not necessary. Like liquid crystals, leucodyes also have a layered molecular structure that changes its orientation due to the changes in temperature.

Today, thermochromic materials are mainly used for making: mugs that change color based on the temperature of the liquid they contain; pens that change color based on one's finger temperature; bath toys that predict overall hotness or coolness of bath water; and clothing whose colors change according to the amount of heat generated by the body. In architecture, these materials are mostly used for interactive visual effects, although they could have additional applications for green architecture in the future.

Thermochromic windows   In response to changes in the ambient temperature, new thermochromic window films alter their color structures as well as reduce solar heat transmission by blocking UV radiation (see Fig. 7.1). These thin plastic films are quite practical and can be incorporated into almost any window assembly.

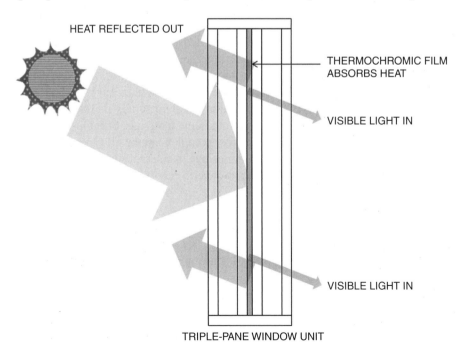

HEAT REFLECTED OUT

THERMOCHROMIC FILM
ABSORBS HEAT

VISIBLE LIGHT IN

VISIBLE LIGHT IN

TRIPLE-PANE WINDOW UNIT

Figure 7.1   Thermochromic windows provide energy savings up to 30 percent over traditional window systems.

For example, sunlight responsive thermochromic (SRT) system is developing a thermochromic film that darkens as it absorbs sunlight shining directly on a window. The tinted film has a low E layer, preventing an undesirable heat load on the building.

Thermochromic paints Thermochromic paints change color (and thus heat absorption) based on temperature changes in the outdoor environment. According to experimental studies done by the Group of Buildings Environmental Studies of the Physics Department at the University of Athens (Karlessia, Santamourisa, Apostolakisb, Synnefaa and Livada 2009) thermochromic applications provide thermally comfortable living environments that reduce energy consumption and help improve the urban microclimate.

THERMOTROPIC

In response to heat and temperature changes, thermotropic materials undergo various property transformations, including: conductivity, transmissivity, volumetric expansion, and solubility (Addington and Schodek 2005). When a thermotropic material absorbs heat its molecular structure is altered, resulting in property changes.

Although they all have similar structures, there are several different thermotropic systems, including casting resins, polymer films, hydrogels, and liquid crystals. Transparent casting resins are used in diverse types of glazing, such as windows, facade elements, and roofs. Thermoplastic polymers are used for existing windows, roof structures, or as laminates for greenhouses and solar panels. Hydrogels are particularly ideal for energy-efficient windows, as they may be used over a wide temperature range (5–60°C) (Schwartz 2008). Thermotropic liquid crystals are similar to liquids and crystalline solids. They are used for smart, efficient windows as they can provide privacy without sacrificing incoming light.

Thermotropic windows Thermotropic windows are one of the most popular smart window systems today because a window's visibility is directly controlled by climatic temperature changes. If the material's temperature exceeds a certain limit, the thermotropic layer becomes milky white, reflecting a large proportion of the incident light. However, there are no visual changes to the window at low temperatures. Therefore, during the winter these windows allow solar light and heat to penetrate undiminished into the building. For more information, see Fraunhofer website: http://www.iap.fraunhofer.de.

THERMOELECTRIC

Thermoelectric is the conversion of electrical energy into thermal energy, which is based on a principle called the "Peltier" or "Peltier–Seebeck" effect (see Fig. 7.2). The principle of thermoelectric energy conversion was discovered by Seebeck in 1821 (Ohta 2007). Seebeck found that an electrical volt was generated between two ends of a metal bar when there was a temperature difference at each end of the bar. In 1834,

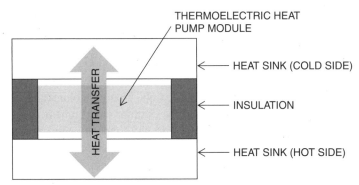

THERMOELECTRIC HEAT
PUMP MODULE

HEAT TRANSFER

←——— HEAT SINK (COLD SIDE)

←——— INSULATION

←——— HEAT SINK (HOT SIDE)

Figure 7.2   A thermoelectric module is a small, light-weight, and silent solid state device that can operate as a heat pump or as an electrical power generator with no moving parts.

Jean Peltier discovered that heat transfers from one metal to another occurred when an electrical current was applied across the junction of two dissimilar metals.

Thermoelectric materials are constructed from a series of connected metals. When an electrical current passes through the connections, heat is transferred. In fact, these materials are capable of transferring a large quantity of heat when connected to a heat-absorbing device on one side and a heat-dissipating device on the other.

There are three physical properties that thermoelectric materials need to efficiently turn thermal energy into electric current:

- Low thermal conductivity
- High electrical conductivity
- Large thermoelectromotive force

Thermoelectric modules are durable, reliable, silent, lightweight, and compact green materials; they do not include compressed gasses, chemicals, or toxic agents. Currently, because of their relatively high cost and low efficiency, thermoelectric devices are limited to applications in which portability, reliability, and/or small size are more important than their cost—such as for the military or aerospace industry.

SHAPE MEMORY

Shape memory materials change their shapes from a rigid form to an elastic state when thermal energy is applied (see Fig. 7.3). When a thermal stimulus is removed, the material reverts back to its original rigid state without degradation. These effects are called "thermal shape memory" and "superelasticity" (Duerig 1990; Lagoudas 2008).

There are two classes of shape memory materials, each with different shape-changing characteristics: shape memory polymers (SMP) and shape memory alloys (SMA). When exposed to a temperature change, SMPs exhibit a mechanical property loss, as seen in releasable fasteners. By contrast, materials with SMAs provide force. As such,

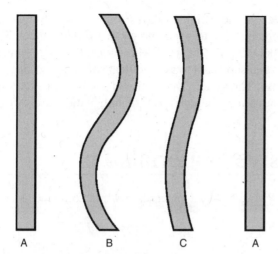

A   B   C   A

Figure 7.3 Shape memory materials
undergo reversible shape changes when
thermal energy is applied.

they can become a lightweight, solid-state alternative to conventional actuators, such
as hydraulic, pneumatic, and motor-based systems.

The application of shape memory materials spans a wide variety of sectors (automotive, biomedical, aerospace, robotics) where their superelastic properties or the
shape memory effect can be best utilized. In green architecture, applications vary from
self-repairing concrete beams (Kuang and Ou 2007) to deployable structures. Practical
applications, such as shape memory foams that expand when exposed to higher temperatures to seal window frames, have been widely used in the building industry.

Self-accommodating ventilation systems This experimental project was
developed by Sergem Engineering BV, located in Leidschendam, Netherlands. In this
project, shape memory alloys are used in a self-accommodating ventilation system.
Changes in temperature within or outside the building activate the SMAs, operating a
louver ventilation system.

Deployable structures A shape memory composite is a unique material for use in
dynamic structures and other applications requiring both load strength and flexibility.

Cornerstone Research Group, Inc., (CRG) has developed a structure system using
shape memory composites that can be temporarily softened, reshaped, and hardened
to function as deployable, flexible structures. These materials can be easily stowed for
space efficiency, and then later deployed to its operational shape. The company states
that the composites can be fabricated from nearly any fiber type, and that creative reinforcements permit dramatic shape changes. Applications include lightweight, rigid
deployable structures (especially as an alternative or enhancement to current inflatable
structures, rapid manufacturing, and dynamic reinforcement).

Flexible surface materials   In their Living Glass Project, David Benjamin and Soo-In Yang used a shape memory alloy to open and close a surface, much like the gills of a fish (Brownell 2006; Manfra 2006; McMasters 2006). Wire embedded within cast silicone contract when subjected to an electrical stimulus, causing "gills" in the surface to open and close. The system can be used for environmental control, such as circulating air within a room when high levels of carbon dioxide are detected.

Light-Responsive Materials

Light-responsive materials (LRMs) are smart materials that transform due to a change in light.

PHOTOCHROMIC

Photochromic materials change their ability to reflect color when exposed to light, and the color change is proportional to its level of UV light absorption. When a photochromic material absorbs UV light, the chemical structure of its molecules (and consequently, the light reflection of the material) changes. This results in reversible color reflections. When the light source is terminated, the material changes back to a clear state. Photochromic materials can also change from one color to another when it is combined with a base color.

Photochromic materials have numerous potential applications, such as for eye glasses, toys, cosmetics, clothing, and industrial products. They are also used in paints, inks, and casting materials. One potential application is on windows for vehicles and buildings, but currently these materials are too unstable, showing weak color change at high temperatures.

PHOTOVOLTAIC

A photovoltaic system is the process of producing an electrical current in a solid material. It was first discovered in 1839, but was not utilized for more than a hundred years. Today, photovoltaic materials are used to convert sunlight into electricity. When a photovoltaic material absorbs UV light, the photons separate the electrons from the atoms of the solar cell material, allowing the free electrons to move through the cell. This creates a new energy that can be harnessed and subsequently used for electrical energy. The physical process of this conversion is called "the photovoltaic effect."

Currently, there are two types of solar cell technologies: (1) crystalline materials and (2) thin film materials. Single crystal silicon cells are made of thin silicon wafers, which are cut from a single silicon crystal. These are the most efficient type of silicon cells and have a long life expectancy of more than 25 years. However, the cells are fragile and must be mounted in a rigid frame. Multicrystal silicon cells are also extremely thin wafers of silicon but are cut from multiple crystals with similar life expectancy and fragility attributes. Multicrystal silicons are slightly less efficient than single crystal cells and require more surface area to produce a given amount of electricity. These types of cells are usually square and have a varied appearance.

Thin film photovoltaic materials produce solar cells with lower conversion efficiencies. They need less direct sunlight and use much less material than crystalline silicon cells (Bube 1998).

Photovoltaic materials have a wide range of application areas, from daily consumer products to hybrid power generators. In architecture, photovoltaic materials are used as custom panels, shingles, solar tiles, and window film applications.

PHOTOLUMINESCENT

Photoluminescent materials absorb radiation from light and convert it into visible light (Addington and Schodek 2005). Photoluminescence materials are often used in our daily life. For example, fluorescent dyes are used for the production of bright textiles, in traffic signs and road markings, in medicine, and in many other science and technology fields. It is especially used for fluorescent lighting, whose tubes include both electroluminescent (mercury) and photoluminescent (fluorescent coating) components. In green architecture, they are widely used for exit signs and other self-luminous emergency egress indicators as they do not rely on external energy sources and require minimal maintenance.

Stimulus (Force)-Responsive Materials

Stimulus (force)-responsive materials (SRMs) are smart materials that transform due to a change in external stimulants, such as electricity, mechanical force, and kinetic energy.

ELECTROCHROMIC

Electrochromic is the ability of a material to transmit light due to a change in electrical current. The optical properties are reversible, and the material reverts to its original state once the electrical current is removed. As such, electrochromic materials are the primary choice for visual devices, such as smart windows, light shutters, information displays, reflectance mirrors, and thermal radiators (Granqvist 1995) (see Fig. 7.4).

Electrochromic materials were first introduced by Deb on tungsten oxide films 35 years ago. (Deb 1973; Granqvist 1995) and became immediately popular for information displays, especially for liquid crystal devices. In green architecture, this material is mainly used in "smart windows" for its energy efficiency and thermal comfort. The transparency/opacity level is adjusted by an applied voltage.

ELECTROSTRICTIVE

Electrostrictive materials change in size in response to an electric field and produce electricity when stretched. When an electrical current is applied, the molecular structure of

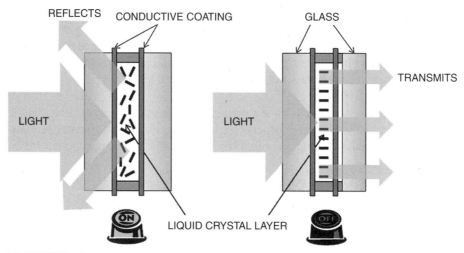

Figure 7.4   **Electrochromic smart windows.**

the material changes through polarization, which alters its molecular energy and produces elastic energy (Addington and Schodek 2005).

These materials are primarily used as precision control systems such as: vibration control and acoustic regulation systems in engineering, vibration damping in floor systems, and dynamic loading in building construction. They are also used as transducers for a variety of electric power generation applications, such as acoustic actuators for smart skins and microactuators for micropumps and valves.

PIEZOELECTRIC

Piezoelectric materials generate an electrical current in response to an applied mechanical stress. Like most electrostrictive materials, piezoelectrics are bidirectional, meaning an applied input produces a deformation. (Addington and Schodek 2005). Piezoelectric materials are used in electromechanical devices such as: microphone transducers, speakers, ceramic tweeters, buzzers, medical ultrasound imaging, and underwater sonar devices.

Nanomaterials

The prefix "nano" is derived from the Greek "nanos," meaning "dwarf" or "little old man." This term is often used to describe a particular length scale, one that typically ranges between 1 and 100 nm. All materials are composed of grains, which in turn comprise many atoms. Nanomaterials, then, are materials possessing grain sizes from 1 to 100 nm in length in at least one coordinate, and often in three coordinates (Wilson, Kannangara, Smith, Simmons and Raguse 2002).

Conventional materials have grains varying in size anywhere from hundreds of microns to centimeters. These grains are usually invisible to the naked eye. In typical nanomaterials, the majority of the atoms are located on the surface of the particles, whereas they are located within the bulk of conventional materials. Thus, the intrinsic properties of nanomaterials are different from conventional materials since the majority of atoms are in a different environment. As such, nanomaterials represent an almost ultimate increasing surface area (Wilson, Kannangara, Smith, Simmons and Raguse 2002). As such, they manifest extremely different and useful properties, which can be utilized for a variety of structural and nonstructural applications.

NANOSCALE STRUCTURES IN UNPROCESSED FORM

Fullerenes   Fullerenes are formal structures of carbon atoms that can be arranged into various geometric shapes, such as cylindrical and spherical. Because of its resemblance, it was named after the geodesic domes designed by the architect Buckminster Fuller (see Fig. 7.5). Different geometrical shapes have different names; for example, spherical fullerenes are called "buckyballs" and cylindrical ones are called "nanotubes."

Fullerenes molecules are similar to graphite and diamond but contain more carbon. This gives them unique structural properties. For example, carbon nanotubes are only a few nanometers in size but 50 percent lighter and 20 times stronger than steel alloys. They have high tensile strength, high resistance to heat, and high electrical conductivity. Because of their excellent nanoproperties, fullerenes are used in various high-end applications such as: sports gear, conductive adhesives, connectors, plastics, molecular electronics, biomedical applications, and for energy storage and thermal materials.

BUCKMINSTER FULLER DOME

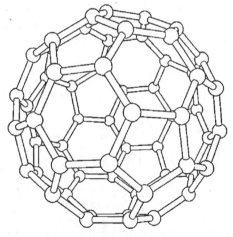

FULLERENES

Figure 7.5   Fullerenes carbon molecules were named after Buckminster Fuller for their resemblance to his geodesic designs.

Nanoparticles Nanoparticles are the building blocks of nanomaterials, with diameters of 100 nm or less. Because of their nanoscale, their properties (mechanical, optical, magnetic, etc.) show molecular differences, and correspondingly, exhibit dissimilar characteristics than the bulk materials. For example, copper nanoparticles do not exhibit deformation characteristics as do bulk copper. Similarly, unlike the bulk counterpart, carbon nanoparticles demonstrate unique properties such as low density, high porosity, improved tensile strength, and thermal and chemical stability. Therefore, they are used in hybrid-specialized applications such as paint, laminating and lubricating agents, electrode materials, and semiconductors. Recent research on nanoparticles has focused on solar energy cells, biomedical applications (e.g., biosensors and imaging devices), and electronics.

Nanowires Also known as "quantum wires," nanowires are ultrathin engineered wires (Wang 2006). Like nanoparticles, nanowires show uniquely enhanced properties over their bulk counterparts. Although most nanowire work is being done in research laboratories today, their incredible length-to-width ratio will allow these nanomaterials to replace nanotubes in the future. Potential uses are for semiconductors, transistors, and biomedical applications. For example, various nanowire-based devices (such as nanobelts and nanosprings) exhibit similar piezoelectric properties and, therefore, can replace piezomaterials for actuators and sensors.

8

CASE STUDIES

This chapter takes an in-depth look at a broad range of relevant case studies in green architecture that are designed, operated, and renovated or reused in an ecological, resource-efficient manner. Selected projects represent both new construction and future projects for various building types, based on the earlier defined standards of green architecture (i.e., employing alternative energies, incorporating advanced technologies and materials, using efficient water and waste management techniques, and reducing the overall impact to the environment). Case studies are presented and analyzed in four major categories.

1 Advanced green buildings
2 Active and passive solar buildings
3 Self-sufficient, off-the-grid modular and mobile systems
4 Solar decathlon competition projects since 2002

Advanced Green Buildings

SAN FRANCISCO FEDERAL BUILDING

- Building type: Government
- Location: San Francisco, California
- Architect: Tom Mayne, Morphosis
- Completion date: 2007

San Francisco Federal Building is a high-performance, green building designed by Morphosis for the General Services Administration (GSA). The building houses various federal offices for the Department of Health and Human Services, Social Security Administration, Department of State, Department of Labor, and the Department of Agriculture. The building program includes 18-story office tower, 3-story elevator atrium lobbies, publicly accessible conference, community center, and day care, fitness center, outdoor sky plaza, and café.

Green features of the building include sunscreens, sensor controlled natural day-lighting, natural ventilation, individually controlled shading devices, flexible floor plans, underground parking, energy-efficient elevators, low- and nontoxic building materials, wood ceilings, and waste management systems. According to the GSA, these features account for 33 percent reduction in energy use, 50 percent blockage in solar radiation, and 7 percent saving in project cost.

Green materials

- Floor to ceiling glass: 13 ft overall ceiling height in the office floors allows light to penetrate deep into work areas, and the occupants to use the natural daylight.
- Fifty percent mix of slag-concrete is used throughout the construction, which makes the concrete durable, stronger, and less pollutant for the environment.
- A mix of white cement used in exposed areas (such as wall and ceilings) makes the utilization of natural light more effective and reduces the energy use.
- Only low- or zero-toxicity building materials (i.e. carpet, paint, furniture) were used.
- During construction, only 10 percent of the materials end up in landfill, and 87 percent of the materials used are recyclable.
- Special dry-polish concrete system, called the Natural Wonder Floor System™, is used as the floor covering material, which is more durable and cost-effective than traditional materials.
- Wood ceilings: Low-emission rate; contains recycled materials; wood comes from a "well-managed" forest; FSC certified.
- Permeable, decomposed granite: Plaza is paved with a permeable, decomposed granite surface that allows rainwater to percolate back into the ground.

Figure 8.1   Wall detail and side view of San Francisco Federal Building.

Figure 8.2   Perforated stainless-steel panels and street view of San Francisco Federal Building.

Green technologies

- Sensors: Series of sensors are embedded in the building to automate temperature and light adjustments.
- Temperature sensors on the facade open the ventilation windows at night and let the outside air get into the building.
- Light sensors detect light and occupancy levels, adjust the lights accordingly. For example, the lights are automatically turned off when the work area is unoccupied.
- Perforated stainless-steel panels: This is a multipurpose panel technology, which serves as a sunscreen and natural ventilation system. As a sunscreen, the panels shade the building from low winter sun and cut daylight to a comfortable level for the workers. As a ventilation system, the panels create a passive heat pump, which

adjusts the heating/cooling load of the building. When air warms up and heat the building, the panels float upward and let cooler air through the building/draw the hot air out of the building by switching the sensors on and automatically opening the windows.

- Vertical glass louvers: Designed to block out low sun angles, and to protect occupants from late-afternoon summer sun.
- Passive design technologies: The narrow shape, the orientation of the building, and the design of the workspaces close to the windows allow the building to absorb maximum amount of daylight, and to utilize the natural ventilation throughout the building. As a result, 80 percent of the workspace is illuminated with natural light, and the top 13 floors of the building do not have air-conditioning.
- Dual flush toilets and drip irrigation: Reduce drinkable water consumption by over 30 percent.
- Acoustical wall technologies are used to absorb the sounds of conversation and background noise.
- Elevators top on every third floor: Forces about 60 percent of tenants to walk up or down a floor to get to their destination; this promotes employees' health and interaction between them. Since there are fewer stops, service is faster and consumes less energy than a traditional elevator.
- Underground parking: Eliminates the waste of unused space, reduces the heat surrounding the building and the city.

BAHRAIN WORLD TRADE CENTER

- Building type: Commercial
- Location: Manama, Bahrain
- Architect: Atkins Architects
- Completion date: 2006

The Bahrain World Trade Center (BWTC) is a high-rise twin tower complex with a rich, mixed-use program, including two 34-story office towers, hotel, shopping mall, and several leisure and dining facilities. The sail-shaped building is 50-story high (787 ft), and features an array of advanced green and smart technologies. The towers are powered by three large wind turbines, while the thick gravel roof works to cool the building. The turbine blades on Bahrain's iconic landmark are the first in the world to be integrated on such a scale into a commercial development. BWTC establishes a technological precedent, which is set to raise the awareness of environmental design and its importance in the built environment. The building paves the way for designers and clients to incorporate renewables and energy-efficient measures to reduce carbon emissions into their future developments.

Green materials

- Thermal glass
- Hundred percent postconsumer recycled materials

Figure 8.3   Two views of the Bahrain World Trade Center. *(Images courtesy of Atkins)*

Green technologies

- Turbines: The BWTC is designed angularly to accelerate and funnel the wind into its three massive (29 m/95 ft diameter) turbines. These turbines provide 11 to 15 percent of the building's electricity.

- Gravel roof: The deep gravel roof works like a traditional green roof to reduce heat. It also provides kinetic insulation.
- Sunshading: The balconies are cantilevered, creating effective sunshades, which reduces heat/solar gain.

QUEENS UNIVERSITY BEAMISH-MUNRO HALL

- Building type: Education/Public
- Location: Kingston, Ontario
- Architect: B+H Architects
- Completion date: March 2004

Beamish-Munro Hall is an academic building that serves as the Integrated Learning Center at Queen's University in Kingston, Ontario. This three-story building contains 100,000 ft^2 of space for faculty, students, and those interested in learning the principles of sustainable buildings. The building's program includes faculty offices, design studios, group rooms, multimedia facility, site investigation facility, active learning center, and various lab facilities.

In conjunction with its educational pedagogy, BMH's design includes advanced green features, including their signature green wall, energy saving systems, smart lighting fixtures, radiant flooring, mechanical systems and ventilation, water conservation fixtures, energy-efficient building envelope, green materials, and passive solar principles.

Green materials
- Concrete with 50 percent recycled supplementary cementing materials (fly ash)
- Wet wall material on the green wall, which filters the contaminants in the air
- Radiant flooring
- Photovoltaic panels
- Fluorescent HIDs
- Scratch-resistant tempered glass
- Ordinary, epoxy-coated, and galvanized rebar as shear wall reinforcement
- Fiber reinforced polymer (FRP) bars are used as slab reinforcement

Green technologies
- Green wall: A three-story green wall filters particles from the air before they are distributed throughout the building.
- High-efficiency lighting systems: High-efficiency lamps and luminaries with electronic ballasts, building automation systems (BAS).
- Intranet controlled lighting.
- Photovoltaic technologies for alternative energy generation.
- Steam heating from the central plant.
- Heat recovery wheel.

Figure 8.4 **Queens University Beamish-Munro Hall.** *(Images courtesy of B+H Architects (left) and Richard Johnson (right and bottom))*

- Displacement ventilation, variable speed fans, and high-efficiency motor.
- Water conservation fixtures (low-flush toilets, proximity detectors for the urinals and faucets).
- Roof storm water collector.
- High-performance building envelope (glazing with a very low U-factor, high-density insulation, and continuous air barrier).
- Acoustic control systems.
- Glass "curtain wall" facing west so students can study how different coating on the windows can prevent heat loss.
- Ceiling painted to reflect light around the room, thus decreasing the use of electricity.
- Sensors in the walls demonstrate how insulation works and how buildings leak and retain heat.
- Optimized passive solar technologies (daylighting for spaces and interior plazas, roof overhang on the west side to reduce solar gain and air-conditioning).

CIS TOWER

- Building type: Commercial
- Location: Manchester, United Kingdom
- Architect: Gordon Tait
- Completion date: 2005

Originally built in 1962, the CIS Tower, at 400 ft, is the second tallest building in Manchester. It is home to Cooperative Financial Services, the UK's largest consumer cooperative. It has recently gained attention by installing a massive solar panel system onto the existing facade. In December 2004, in a 10 million dollar project, Gordon Tait as the leading architect redesigned the CIS Tower's concrete service tower by adding a new cladding technology. This new photovoltaic cladding replaced the old, deteriorating exterior of mosaic tiles, which were falling off the face of the building and presented a safety hazard.

This massive solar panel system now provides 10 percent of the building's electricity. The solar paneling is the UK's largest solar array and Europe's largest vertical solar array. Solar panels generate 181,000 units of renewable electricity each year, equivalent to the energy needed to power 55 homes for a year. Its best performance statistic is its annual saving of over 100 tons of CO_2 emissions.

Green materials   No green materials were used in the construction of this building when it was originally built in 1962. Standard construction practices and materials were used at the time.

Green technologies
- Three sides of the service tower are wrapped with photovoltaic panels.
- The photovoltaic panels provide protection from the elements. They reduce the heating and cooling load for the building.

■ Seven thousand two hundred and forty four 80-W photovoltaic panels: Generate 180,000 kWh of renewable energy per year and provide 10 percent of the building's energy needs. However, only 4898 modules are actually in operation; the rest are aesthetic modules to match the rest of the facade.

DES MOINES PUBLIC LIBRARY

■ Building type: Public/Education
■ Location: Des Moines, Iowa
■ Architect: David Chipperfield Architects
■ Completion date: 2006

The Des Moines Public Library (DMPL) is a green building with a public-oriented program, which includes education facilities, children's play areas, book stacks, conference hall, café, administration, storage, underground parking, and service areas. Designed by David Chipperfield Architects and completed in 2006, the building promotes transparency, openness to the public, and sustainability.

The green features of the two-story concrete building are a copper-glass skin, a green roof, passive solar technologies, and exposed concrete soffits. The successful design of DMPL helps in reducing solar gain and keeps it naturally cool in the summer. The cumulative effect of these measures have led to an excellent energy rating with the local energy supplier and given the library a significant cost rebate in addition to annual savings on their energy bill.

Green materials

■ Exposed concrete soffits: All suspended ceilings have been omitted, exposing the concrete soffit of the floor slabs. As the slabs are exposed, their building mass can be activated to reduce the building's cooling load.
■ Copper-glass skin: A recyclable material with a long life span, the three-dimensional quality of the mesh allows for good views from the building but reduces solar gain through the façade by 80 percent, thus significantly reducing the building's cooling load.

Green technologies

■ Green roof: The green roof acts as a natural thermal shield and helps maintain the temperature of the building. It also slows down rain, rather than shedding it into the environment and flooding sewer systems. The roof is made of many other layers besides the layer of sedum and plants or the fabric layer which stops roots. Below these layers are: a layer for water storage, aeration, and drainage; a layer of polystyrene insulation; a layer protecting against water damage, UV light, root growth and chemicals, all of which rest on the actual concrete roof.

Figure 8.5   Des Moines Public Library (top): copper exterior (center) and interior detail (bottom). *(Images courtesy of David Chipperfield Architects and photographer Farshid Assassi)*

■ Passive solar technologies: The elaborate building shape helps to connect inside and outside, and maximizes the use of natural daylight. Integrating mesh in the full-height glazing furthers the effectiveness of the daylight. The mesh mitigates the sometimes harsh qualities of daylight, thus minimizing the use of artificial light to avoid contrast and helping to illuminate the entire depth of the space.

Figure 8.5   *(Continued)*

EDITT TOWER

- Building type: Commercial
- Location: Singapore
- Architect: TR Hamzah & Ken Yeang
- Completion date: Construction pending

EDITT Tower is an exposition tower competition project with a still-pending competition date. However, the green features of this project are state of the art and so unique that it is worth covering the project in this book. EDITT Tower is primarily designed to demonstrate an ecological approach to tower construction. Besides meeting the program requirements of the client for an exposition tower (i.e., of retail, exhibition spaces, auditorium uses, etc.), the design follows many green approaches to the site: place making (creating a pleasing, interesting landmark), vertical landscaping, water and waste management, solar energy use, storm water collection, green materials and technologies, and concern for environmental health.

The 26-story high-rise will boast alternative energy, natural ventilation, and a biogas generation plant, all wrapped within an insulating living wall that covers half of its surface area. The ecoskyscraper is designed to enrich the biodiversity of the area and support the local ecosystem. EDITT Tower's other green feature includes future adaptability for future uses.

Green materials
- Planted facades and vegetated-terraces..
- Recyclable materials: The design has a built-in waste-management system. Recyclable materials are separated at source by hoppers at every floor. The materials drop down

to the basement waste-separators, where they are kept for collection by recycling garbage collection.

- Low-embodied energy materials: Embodied-energy studies of the building are useful to indicate the building's environmental impacts. Subsequently, estimates of CO_2 emissions arising from building materials production may be made. High embodied–energy materials (e.g., aluminum and steel) will be recycled, thereby halving their embodied energy when reused. Replacing concrete floors with composite-timber floors will reduce embodied energy by approximately 10,000 GJ.

Green technologies

- Stormwater collection system: Water self-sufficiency (by rainwater collection and gray water reuse) in the tower is at 55.1 percent.
- Water purification technology: The rainwater collection system is comprised of "roof-catchment-pan" and layers of "scallops" located at the building's facade to catch rainwater running off its sides. Water flows through a gravity-fed water-purification system, using soil-bed filters. The filtered water accumulates in a basement storage tank, and is pumped to the upper-level storage tank for reuse (e.g., for plant irrigation and toilet-flushing).
- Sewage recycling: Sewage is treated to create compost (fertilizer for use elsewhere) or biogas fuel.
- Solar energy use: Photovoltaic technologies are used for greater energy self-sufficiency.

Figure 8.6   Five views of the EDITT Tower with graphic specifications.

(Images courtesy of Ken Yeang)

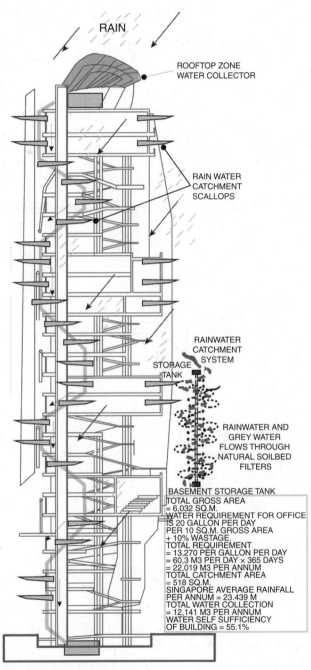

RAIN

ROOFTOP ZONE
WATER COLLECTOR

RAIN WATER
CATCHMENT
SCALLOPS

RAINWATER
CATCHMENT
SYSTEM

STORAGE
TANK

RAINWATER AND
GREY WATER
FLOWS THROUGH
NATURAL SOILBED
FILTERS

BASEMENT STORAGE TANK
TOTAL GROSS AREA
= 6,032 SQ.M.
WATER REQUIREMENT FOR OFFICE
IS 20 GALLON PER DAY
PER 10 SQ.M. GROSS AREA
+ 10% WASTAGE.
TOTAL REQUIREMENT
= 13,270 PER GALLON PER DAY
= 60,3 M3 PER DAY × 365 DAYS
= 22,019 M3 PER ANNUM
TOTAL CATCHMENT AREA
= 518 SQ.M.
SINGAPORE AVERAGE RAINFALL
PER ANNUM = 23.439 M
TOTAL WATER COLLECTION
= 12,141 M3 PER ANNUM
WATER SELF SUFFICIENCY
OF BUILDING = 55.1%

○ **RAINWATER COLLECTION AND
RECYCLING SYSTEM**

Figure 8.6   *(Continued)*

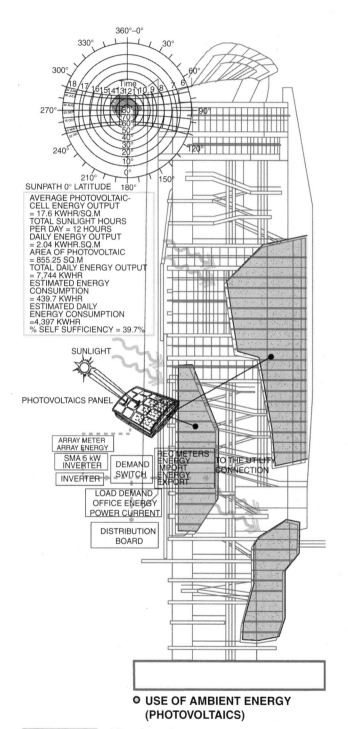

360°–0°
330°   30°
300°   60°
                      18 17 16 15 14 13 12 11 10 9 8 7 6
270°   Time   90°
                180°
                170°
240°   160°   120°
                150°
                140°
                130°
                120°
                110°
                100°
                90°
                80°
                70°
                60°
                50°
                40°
                30°
                20°
                10°
                0°
210°   180°   150°

22 JUN
30 JUL
30 AUG
23 SEP
15 OCT
22 NOV
22 DEC

SUNPATH 0° LATITUDE

AVERAGE PHOTOVOLTAIC-
CELL ENERGY OUTPUT
= 17.6 KWHR/SQ.M
TOTAL SUNLIGHT HOURS
PER DAY = 12 HOURS
DAILY ENERGY OUTPUT
= 2.04 KWHR.SQ.M
AREA OF PHOTOVOLTAIC
= 855.25 SQ.M
TOTAL DAILY ENERGY OUTPUT
= 7,744 KWHR
ESTIMATED ENERGY
CONSUMPTION
= 439.7 KWHR
ESTIMATED DAILY
ENERGY CONSUMPTION
=4,397 KWHR
% SELF SUFFICIENCY = 39.7%

SUNLIGHT

PHOTOVOLTAICS PANEL

ARRAY METER
ARRAY ENERGY

SMA 5 kW
INVERTER

INVERTER

DEMAND
SWITCH

REC METERS
ENERGY
IMPORT
ENERGY
EXPORT

TO THE UTILITY
CONNECTION

LOAD DEMAND
OFFICE ENERGY
POWER CURRENT

DISTRIBUTION
BOARD

○ **USE OF AMBIENT ENERGY
(PHOTOVOLTAICS)**

Figure 8.6   (*Continued*)

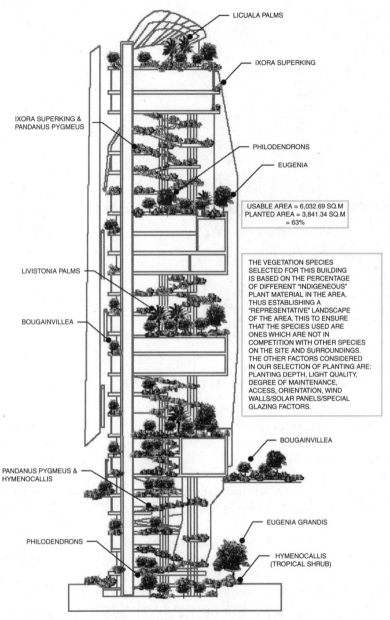

LICUALA PALMS

IXORA SUPERKING

IXORA SUPERKING &
PANDANUS PYGMEUS

PHILODENDRONS

EUGENIA

USABLE AREA = 6,032.69 SQ.M
PLANTED AREA = 3,841.34 SQ.M
= 63%

THE VEGETATION SPECIES
SELECTED FOR THIS BUILDING
IS BASED ON THE PERCENTAGE
OF DIFFERENT "INDIGENEOUS"
PLANT MATERIAL IN THE AREA,
THUS ESTABLISHING A
"REPRESENTATIVE" LANDSCAPE
OF THE AREA. THIS TO ENSURE
THAT THE SPECIES USED ARE
ONES WHICH ARE NOT IN
COMPETITION WITH OTHER SPECIES
ON THE SITE AND SURROUNDINGS.
THE OTHER FACTORS CONSIDERED
IN OUR SELECTION OF PLANTING ARE:
PLANTING DEPTH, LIGHT QUALITY,
DEGREE OF MAINTENANCE,
ACCESS, ORIENTATION, WIND
WALLS/SOLAR PANELS/SPECIAL
GLAZING FACTORS.

LIVISTONIA PALMS

BOUGAINVILLEA

BOUGAINVILLEA

PANDANUS PYGMEUS &
HYMENOCALLIS

EUGENIA GRANDIS

HYMENOCALLIS
(TROPICAL SHRUB)

PHILODENDRONS

**o REINTRODUCTION OF ORGANIC MASS
TO URBAN SITE TO COUNTERBALANCE
INORGANIC NATURE OF THE SITE**

Figure 8.6   *(Continued)*

Figure 8.6   (*Continued*)

GOVERNMENT CANYON VISITOR CENTER

- Building type: Public/Commercial
- Location: Helotes, Texas
- Architect: Lake Flato Architects
- Completion date: 2005

The Government Canyon Visitor Center is a high-performance green building designed by Lake Flato Architects in San Antonio, Texas. The visitor center forms the gateway to the 8600 acre Government Canyon State Natural area. The building's program includes an exhibit hall, a park store, classrooms, offices, and an outdoor pavilion. The goal of the project was to create an aquifer preserve in order to protect and restore the natural landscape while creating high-use, low-maintenance, and economical structures that reinforce the mission of the natural area. Development concentrated on reducing landscape water usage and the overall physical impact on the site.

Green features of the building include a narrow floor plate, combined with deep porches, large overhanging roofs, high performance glazing, and reflective roofing. All these features minimize the cooling loads, while allowing daylight to penetrate into interior spaces whenever possible. Photo sensors and occupancy-sensor controls, combined with effective daylighting and energy-efficient fixtures, all contribute to reduce the amount of energy consumed by the building. "Gravity flow water systems, coupled with solar-powered water pumps, efficiently convey water while demonstrating renewable-energy technology."

Green materials

- Rusted steel pipe with a minimum of 75 percent recycled content was extensively used throughout the project.
- The use of naturally oxidized surfaces eliminates the need for paint, thus minimizing the need for maintenance.
- Exposed native stone walls and stone fences on site reflect the history of the place.
- Fly ash concrete provides long life for the building with very low maintenance.
- Corrugated, galvanized metal roofing provides reflective surfaces while reducing the need for roof decking and substructure.
- Eastern red cedar siding, harvested locally, is naturally resistant to decay.
- Recycled-cotton insulation helps to keep heat inside in an ecofriendly way.
- Three main aspects of materials were taken into consideration when making decisions: the durability, maintenance, and regional availability.
- Bronze screens allow breezes to naturally cool the main exhibit space.

Green technologies

- Passive technologies: "The narrow building footprint allows for maximum use of indirect light from the north and south in all occupied spaces. Approximately 90 percent of occupied spaces enjoy effective daylight and views, and 100 percent of spaces have ventilation control."
- Dimmer switches are used in exhibit spaces to make use of both daylight and small amounts of electric lighting when needed.
- All windows are fully operable and oriented to catch both direct light and cooling breezes. "Exhibit and circulation spaces were designed as sheltered and shaded outdoor spaces...accepting summer breezes but protected from north winds."
- In an effort to conserve both materials and energy, many spaces within the building are not air-conditioned (reducing air-conditioned spaces by 35 percent).
- "Extensive use of conventional double-hung windows maximizes the open area and minimizes interference with work paths, ensuring optimal usability."
- Massive rolling doors are used to protect from harsh winds, while also allowing spaces to be fully open to air and light when desirable.
- Rainwater is collected from the roof and used for landscape irrigation and wastewater conveyance. Excess water is stored in partially exposed, underground concrete systems.
- "Solar-powered pumps lift the stored groundwater to the storage tank tower above, providing gravity feed for drip irrigation."

HEIFER INTERNATIONAL CENTER

- Building type: Commercial
- Location: Little Rock, Arkansas
- Architect: Reese Rowland, Polk Stanley
- Completion date: 2006

Figure 8.7   Four views of the Government Canyon Visitor Center. *(Images courtesy of Chris Cooper Photography)*

The Heifer International Center is the building headquarters for the nonprofit organization Heifer International. This green building was designed by architect Reese Rowland to house the company's 450 staff members and to provide public space for their educational and outreach programs. The building was completed by CDI Contractors, LLC of Little Rock, in 2006.

Green features are integrated throughout the building, including recycled materials, innovative green technologies, and sustainable landscape design. Some green aspects include water reclamation features, a water conservation system and water tower, raised

Figure 8.7   *(Continued)*

flooring, passive solar lighting, and reused materials, including a high content of recy-cled steel from the abandoned buildings on the site. Overall, this office building uses less than 50 percent of the energy of a traditional office building of this size, and is pro-jected to last for at least 100 years.

Green materials

- Permeable paving system: This encourages storm water infiltration, which is very ecological. Traditional paving systems drain on-site water at a rate faster than the historic average.
- Chillers: Chillers remove heat from the water surrounding the building on three sides through an "absorption refrigeration cycle." Chilled water is then used to dehumidify and cool the air in the building.
- Recycled-content steel: This reduces manufacturing impact because existing steel is reused and it is still very durable. Also it can be reused when deconstructed.
- Low-VOC materials were selected for the interior finishes, which promote a clean working environment.
- Recycled carpeting: Less earth resources are used to create new carpet.
- Bamboo flooring: A very renewable wood source because it grows so fast and is a biodegradable material.
- Glass skin: Gives inhabitants natural daylight, cutting down on electricity con-sumption.

- Raised flooring: Beneath the raised floor run energy-saving ductwork, modular electrical conduits, and mechanical systems.
- Local materials with low EE: A high percentage of building materials were brought from within 500 miles of the site from steel and aluminum manufacturing facilities in Little Rock.

Green technologies

- Glass staircases placed on exterior: Architect Rowland says, "The staircases promote health and they cut down on elevator use." They also allow natural light to flood in.
- Daylighting: The glass exterior lets in plenty of natural light and reduces the energy to power the building.
- Passive solar: Overhangs reduce unwanted solar heat gain.
- Roof water collection: A roof tower collects and recycles graywater for flushing toilets and urinals, and radiant heating system.
- Gray water storage tank: The gray water is collected from the roof, from the lavatories, and from the condensate from the ventilating units. The gray water supplies 90 percent of the building's water needs for the toilets and the cooling tower.
- Post project completion maintenance: A series of check valves were replaced on the HVAC system after it was discovered that $20,000 a year would have been wasted by the initial valves' inefficiency. Also, there were leaks around data boxes in the access flooring that caused unnecessary energy loss. The under-floor air distribution system had to be tightened up.
- Occupancy sensors: These reduce lighting use and cooling load, maximizing energy efficiency.

MAIN STATION

- Building type: Public/Commercial
- Location: Stuttgart, Germany
- Architect: Ingenhoven Architects
- Completion date: 2013

The project as envisioned will transform Stuttgart's nineteenth-century railroad terminal into a modern, regional transit hub. More importantly, the design team focused on creating new materials and technologies to make this a zero-energy building as well as integrating it into the surrounding community, which consists of a park and several municipal buildings in a downtown urban scene.

The Main Station is quite unique, and though we will not truly learn its success as a green building until it is built and functioning, it can be evaluated now on the strong claims it proposes. The main components of the underground Main Station are an expansive, durable, tent-like structure made of concrete with about 30 light shafts protruding through the concrete and into the park space above. These light shafts would provide extensive daylight and ventilation. It remains to be seen whether this plan will work properly when built; however, it is an ideal concept for a green building. All in all, it will be sustainable for a very long time and use very little energy.

Green materials

- Continuous form of shell-type concrete: Extremely strong and sustains longer than other forms of concrete. Concrete itself is composed of materials that make it extremely strong; in this continuous form it is still more efficient structurally. The mysterious components in the concrete have not been revealed at this time.
- Reflective ceiling material is used to reflect the artificial light at nighttime, reducing the total lighting load, which lengthens the life of the lighting system and saves energy.
- Light-colored surfaces are used in the interior to reflect light during the day and night, as well as reduce the need for artificial lighting, again saving energy.
- Solar panels located on the north side of the building produce enough energy to sustain the artificial lighting needed, which reduces CO_2 emissions significantly.

Green technologies

- The shape of the light shafts allows for an abundance of natural lighting to enter the building throughout the day, which saves energy and CO_2 emissions.
- Natural ventilation is available through apertures in the light shafts, saving energy through less venting loads and improving air quality.
- The shape of the concrete allows the form to cover the entire space, which is very unusual for concrete. This shape greatly assists in the strength and durability of the material.

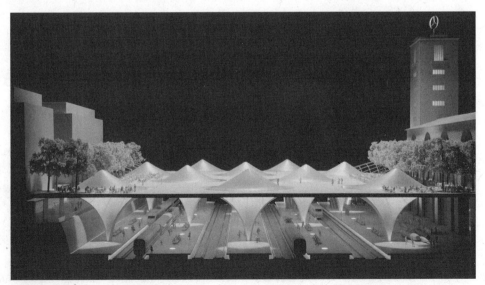

Figure 8.8   **Main Station.** *(Images courtesy of Holger Knauf (top and center) and Peter Wels (bottom))*

Figure 8.8   (*Continued*)

MANITOBA HYDRO HEADQUARTERS

- Building type: Commercial
- Location: Winnipeg, Manitoba (Canada)
- Architect: KPMB Architects and Smith Carter Architects
- Completion date: 2008

Manitoba Hydro Headquarters is located in one of the most extreme climates, varying 70°C over a year and experiencing unusually strong southerly winds and an abundance of sunshine. This provided a huge challenge for the architects who were assigned to create an advanced green building in Winnipeg. The company, one of the largest electric utility companies in Canada, wished to design a building that would demonstrate Manitoba Hydro's ability to cut energy expenditure by 60 percent. The building incorporates green design principles to overcome the harsh climate in several ways. For example, the form and orientation of the building are designed to maximize passive solar energy, natural ventilation, heating, and cooling. Partitioned double walls in between towers act as solar collectors, air exchangers, and air shafts. Combined with a solar chimney, the atrium provides natural ventilation throughout the building.

Green materials

- A living "green" roof with mosses, grasses, and lichens provides habitat for local species and helps reduce the urban heat-island effect.
- Atriums provide conditioned fresh air to create a more work-conducive space for employees.
- Structural and glazing systems emphasize lightness and transparency to maximize natural daylight, thereby cutting back on artificial lighting costs.

Green technologies

- To regulate the humidity, 300 flat Mylar cables hang six stories, from ceiling to floor, with water running along them. In the winter, water vapor from the cables humidifies the air. The cables cool water in the summer to 15°C, making the water vapor condense on the cables, which removes the humidity. The cables also act as an interesting art feature.
- A double external wall reduces heating and cooling requirements in extreme temperatures.
- Two towers converge at the north and splay open to the south to capture strong winds and sunlight, utilizing earth's natural resources.
- North and south stacked atria fuse the masses of the building together at each end and function as solar collectors, air exchangers, handlers, and shafts.
- A geothermal heat pump system extracts heat from the ground in the winter and returns it to the ground in warmer temperatures.
- A thermal chimney/solar chimney (covers all 22 floors) promotes natural ventilation, providing the inhabitants with 100 percent fresh outdoor air. It improves the natural ventilation by using convection or air heated by passive solar energy. Solar energy heats the chimney and the air within it, creating an updraft of air. The suction created

Figure 8.9 Two external views and one internal view of the Manitoba Hydro Headquarters. *(Images courtesy of Kuwabara Payne McKenna Blumberg Architects, Design Architects)*

at the base can be used to ventilate and cool the building. The chimney humidifies the building in winter and dehumidifies the building in summer. It can also utilize wind power in addition to solar.

QUEENS BOTANICAL GARDEN VISITOR & ADMINISTRATION CENTER

- Building type: Commercial
- Location: Flushing, New York
- Architect: BKSK Architects
- Completion date: 2007

An extension of the Queens Botanical Garden, the Visitor and Administration Center demonstrates environmental sustainability while incorporating the Garden's mission of connecting people with plants. The building is composed of three interconnected program spaces: a forecourt and roof canopy; a central reception and administration center; and an auditorium blended into the landscape. Green elements include a geothermal system, photovoltaic panels, a stormwater management system, and extensive use of recycled and renewable materials.

Green materials
- Materials used had high durability, low maintenance, recycled content, and low chemical emissions to benefit the environment.
- FSC-certified red-cedar siding and black locust for exterior.
- Exterior covered with brise-soleil (sun blockers or baffles), which are placed on angles based on the latitude of the building; they let sun in during the winter and shade in the summer.
- Green roof above auditorium containing sedum, grasses, and perennial flowers.
- Veneer and bamboo panels which are biodegradable.
- Salvaged hemlock for formwork, eliminating the need for harvesting more trees.
- Use of recycled materials means no production of more products which increase CO_2.

Green technologies
- Geothermal heating/cooling system: Contributes to an overall 41 percent energy cost savings.
- Photovoltaic panels: Generate 17.5 percent of the energy used by the building.
- Wooden brise-soleil: Provides thermal comfort and consistent light levels (on the western and southern wall of the building).
- Reflecting paint on the building facades: Reduces the heat island effect.
- Passive solar technologies: The narrow shape of the building and open plan allow light penetration to the interior (84 percent of the building is daylit).
- Green roof: The building uses the site as a garden wall. The garden roof of the auditorium rises up from the landscape and provides thermal and acoustical insulation.
- Natural ventilation is used in 95 percent of the building.

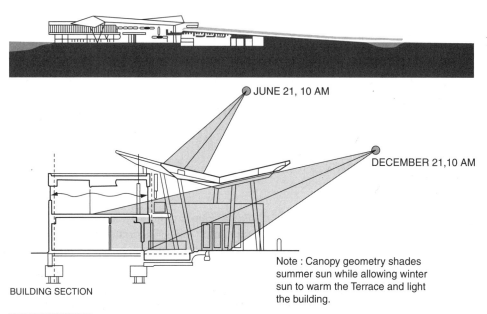

JUNE 21, 10 AM

DECEMBER 21,10 AM

Note : Canopy geometry shades
summer sun while allowing winter
sun to warm the Terrace and light
the building.

BUILDING SECTION

Figure 8.10 **Two views of the Queens Botanical Garden Visitor & Administration Center.** *(Images courtesy of BKSK Architects)*

- Light sensors reduce energy consumption.
- Rainwater collection system: A 24,000-gal cistern stores rainwater for reuse in the water channels and entrance fountain.
- Stormwater management system: All storm water is managed on site. Bioswales (sloped and planted drainage courses) naturally manage stormwater overflow and sustain the surrounding landscape.
- Gray water system: Gray water is used for flushing toilets, reducing water consumption by 55 percent; Waterless urinals reduce water use and conserve the water supply.

THE COUNCIL HOUSE 2 BUILDING

- Building type: Commercial/Public
- Location: Melbourne, Australia
- Architect: Mike Pearce, DesignInc
- Completion date: 2006

Council House 2 (CH2) is a 10-story office building, housing approximately 540 Melbourne city employees, with ground-floor retail spaces and underground parking. It is designed as a pilot project for the City of Melbourne's effort to reduce the CO_2 emission rate to zero by 2020. Among other green technologies, the CH2 building relies on passive energy systems to reduce electricity consumption by 85 percent, gas emissions by 87 percent, water consumption by 72 percent, and CO_2 emissions by

87 percent. Green elements include hydraulically controlled shutters, phase change materials, chilled ceilings, solar-thermal and solar-photovoltaic systems, and a wind turbine.

Green materials

- Hydraulically controlled, recycled timber shutters on the west side automatically open and close depending on the sun's position. This system saves energy through operation by a natural resource and by reducing heating/cooling loads through shading. The timber shutters are recycled, giving them a second life and reducing landfill.
- Phase change materials (PCM) are natural compounds that first collect and then release energy. Here a storage tank made of phase change material is used to re-cool water, which sustains itself through natural resources and saves energy.
- Chilled ceilings are used to absorb radiated heat from equipment and occupants, which is a natural way to cool a space and is fairly low maintenance.
- "Shading and light shelves" are produced by a series of balconies along the north façade that block the high summer sun while providing a light shelf for light to bounce off in the winter. This system saves energy and reduces CO_2 emissions due to reduction in heating/cooling loads as well as lighting loads because sunlight still penetrates through to the interior.
- Thermal mass in concrete slabs is used to absorb excess heat from each space, which reduces cooling loads and thus, saves energy.
- Vertical planting along the north façade and roof provides better air quality and enhances productivity by relieving stress through access to nature. This system also provides shade for the windows.

Green technologies

- Solar-thermal system sustains the energy use of the occupants by heating water using natural resources, reducing CO_2 emissions.
- Solar panel system produces energy, reducing energy use and CO_2 emissions. Healthy air is provided through 100 percent outside air supply ducts and is supplied floor by floor through controllable floor vents, which saves energy by only operating when the space is occupied.
- Shower towers that act as exposed cooling towers for the mechanical system are used as a natural way to cool.
- A wind turbine is located on the roof to provide a negative pressure for the ventilation system, using natural forces to purge air from the building. This system saves energy by not requiring energy-powered pumps and fans.

WHITNEY WATER PURIFICATION FACILITY AND PARK

- Building type: Commercial/Educational
- Location: New Haven, Connecticut
- Architect: Steven Holl Architects
- Completion date: 2005

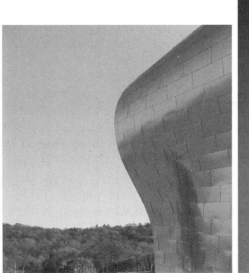

Figure 8.11   Three views of the Whitney Water Purification Facility and Park.
(Images courtesy of Steven Holl Architects)

Designed by the architecture firm of Steven Holl Architects, the Whitney Water Purification Plant provides clean water to the residents of South Central Connecticut. Built to demonstrate the best green design and watershed practices available today, this water purification facility features the largest green roof in Connecticut, zero off-site stormwater discharge, expanded wetlands for biodiversity, and geothermal heating and cooling. Factors determining material choice include durability, recycled content, rapidly renewable content, and low chemical emissions.

Green materials

- Using natural laws, water travels through a series of gardens: Chlorophyll Garden, Wave Meadow, Bubbling Marsh, Filter Court, Wetlands Pond, Turbulence Lawn. The use of local earth resources allows for low emissions rate and biodegradability.
- A stainless steel recyclable skin is an example of effective reuse of material and also aids in reduction of heat absorption.
- Cork tile floor is a composite of recyclable wood content and tree bark. This composite's effective reuse of earth materials allows it to be durable, biodegradable, renewable, and economical—an overall green product.
- The terrazzo floor is a composite material made of recycled glass chip aggregate and Portland cement. This is a very durable and economical example of material recyclability.
- The green roof (30,000 ft^2) and surrounding landscaping eliminate the heat island effect while increasing the insulation R value by 3 points and reducing maintenance and irrigation costs. Native plant species were used on the green roof, and all excavated earth and debris were reused in landscaping and construction. This is a high performance, ecological, and sustainable water purification system.

Green technologies

- The series of gardens filters natural resources in new ways as described above. This system is a smart design utilizing ecological technologies. Earth resources are used efficiently and effectively to purify water for Connecticut.
- The use of a massive array of photovoltaic cells is relatively ecological. There are emissions and waste produced in production and transportation, but after installation there are no emissions or pollution. The smart technologies used are nonrecyclable and nonbiodegradable. Though energy efficient and low maintenance, this technology is only somewhat sustainable due to its limited life span, nonrecyclability, and long-term economic turnover. These solar panels are efficient, effective, and productive, making them a high-performance technology. They are also healthy and safe.

FOUR TIMES SQUARE

- Building type: Commercial
- Location: New York City, New York
- Architect: Fox & Fowle Architects
- Completion date: 1999

Four Times Square (FTS) is a $500 million office tower in Manhattan with an area of 1.6 million ft^2 and 48 stories. FTS was the original skyscraper to incorporate energy efficiency, indoor air quality, sustainable materials, and responsible construction. Operational costs for FTS are 10 to 15 percent lower than other skyscraper projects of this size.

FTS is important as the forerunner of an energy-efficient skyscraper ahead of its time. With two fuel cells and photovoltaic panels, the tower is mostly sustainable. Fifteen kilowatt-hours of energy is generated by the photovoltaic installation, and the fuel cells provide 100 percent of the nighttime energy. FTS is also efficient at air quality control, circulating 50 percent more clean air than city requirements. To avoid street exhaust, outside air enters the building several stories above street level.

Green materials

- Recycled aluminum metal ceiling panels reduce the need for new material and the pollution that mining for that material would produce.
- Sustainable harvested wood used throughout is less polluting and can be fairly easily regrown.
- The hat truss (frame roof system) is built from 65 percent of recycled materials from other deconstruction projects. It uses less material, sustains more weight, and pollutes the environment less.
- Glass curtain walls increase daylight and make the tower nearly electric light free. Low-E glazing helps to prevent direct solar gain.

Green technologies

- Lighting (LED signage, occupancy sensors, fiber optics) will save energy and cut down on day-to-day cost. Occupancy sensors turn lights off when people leave a room.
- Direct-fired absorption chillers and heaters cut down on the total amount of piping used and in three years will pay back the initial cost. Natural gas is used to run the cooling system.
- Fuel cells, used to provide energy at night, are nonpolluting, producing only CO_2 and H_2O as by-products.
- The photovoltaic panels, which produce day-time energy, are placed between windows on the eastern and southern facades near the top of the building.
- Fans/pumps with variable speed and sensors are used to control ventilation.

INFINITY TOWER

- Building type: Commercial
- Location: Dubai, United Arab Emirates
- Architect: Skidmore, Owings & Merrill
- Completion date: Estimated 2010

The Infinity Tower in Dubai is a sleek, twisting design of 73 floors containing 632 flats. The building is functional, while stressing aesthetic appearance and connection to the natural environment. The Infinity Tower will be energy efficient and thermal resistant with respect to the temperature fluctuations in Dubai. Strategies have been used to minimize energy use and reduce the impact of the environment.

The elements used in the Infinity Tower promise to make the building more efficient. Solar panels along the building will produce 25 percent of the energy used during the day. Titanium sheeting and glazed windows will provide light during the day through reflection and direct sunlight. Screen panels will filter out the sun glare for residents. The aluminum framework can be recycled and is lightweight. Although the tower is quite energy efficient on its own, it does not seem to produce additional power for the grid system.

Green materials

- Titanium metal panels used throughout. Increased cost, but require less material to sustain required weight. It has a low density and is a strong, lustrous, corrosion-resistant metal. Can be recycled.
- Aluminum (framework) is used due to its lightweight and high tinsel strength. This metal can be recycled, but if not available it is the most abundant metal on earth.

Green technologies

- Solar panels used to produce energy for day-to-day use. Used to produce pollution-free energy and is completely sustainable.
- Screen panels used to filter out the sunlight in the building. Helps to cut down on greenhouse pollution and keep patrons inside at comfortable temperature levels.
- Glazed windows used to bring in the natural sunlight while cut down on greenhouse pollution.

WIND TUNNEL FOOTBRIDGE

- Building type: Installation/Conceptual
- Location: N/A Prototype
- Architect: Michael Jantzen
- Completion date: N/A

The Wind Tunnel Footbridge is the conceptual energy-efficient bridge. The footbridge will be made of five turbine wheels turning different speeds and different directions. The bridge will produce and store energy for other uses if not for the grid. The Wind Tunnel Footbridge is constructed for various venues as an architectural attraction that produces power.

The Wind Tunnel Footbridge will be wind activated. The turbine wheel will spin and produce different electronic sounds. The electricity will be produced and stored much like a windmill. Any extra power can be put back into the grid. The materials used such as steel and aluminum can be recycled and reused. The Wind Tunnel Footbridge will produce 100 percent of its own energy and is mostly sustainable. The materials, aluminum and steel, will need special weather coating and regular maintenance to be fully sustainable.

Figure 8.12   Three views of the Wind Tunnel Footbridge.

(Images courtesy of Michael Jantzen)

Green materials

■ Steel is used for structural integrity. Has high tinsel strength but weighs a large amount. Will be used from recycled metal, therefore is green.

■ Aluminum (turbine wings) is used due to its lightweight and high tinsel strength. This metal can be recycled, but if not available, it is the most abundant metal on earth.

Green technologies

■ Wind-powered turbines used to produce excess energy. Is a pollution-free way to obtain massive amounts of energy for other uses.

YALE SCULPTURE BUILDING AND GALLERY

■ Building type: Commercial
■ Location: New Haven, Connecticut
■ Architect: Johann Mordhorst with Kieran Timberlake Associates
■ Completion date: 2007

Yale University needed a building to create and display new sculpture and artwork, with classrooms, studios, and administration offices. Stephen Kieran of Timberlake Associates wanted the building's environmental performance to "drive the appearance," and energy conservation and air quality were two primary focus points. The Sculpture Building is a four-story studio-art complex, including an art gallery and an adjacent four-story parking garage with restaurants and retail space.

Green materials

- Cedar cladding, reclaimed from 100-year-old wine barrels, was used in the gallery.
- Porous asphalt is used for walkways to reduce runoff.
- Waterless urinals, low-flow faucets, and dual flush toilets have been installed throughout.
- The curtain or window wall features Nanogel, a new product that comprises the translucent panels. It is able to promote remarkable energy savings.

Green technologies

- More than 89 percent of the waste materials generated during construction were recycled. Seventy eight percent of the materials used during construction were transported from distances within 500 miles.
- Insulation panels: 100 percent recycled newspapers were shredded and blown into wall cavities, promoting sustainability and reducing waste.
- Individual recycling receptacles: Labeled trash bins help keep a cleaner workplace and encourage recycling.
- Dual flush toilets: These toilets provide a full flush for solids and a half flush for liquids. The water comes from a stormwater retention system, designed to reduce annual water use by 65 percent.
- Triple-glazed, argon-filled window panels: Heat-fused glass panes or insulated glass help shut out heat and cold.
- Spandrel panels, which are placed between the window panels, have an insulation value of more than R-20.
- Lighting sensors dim artificial lighting when outside light is sufficient and maintain a constant light level.
- Aircuity air monitoring system: This is a testing device that monitors carbon monoxide, and VOC.
- Curtain wall: The wall provides occupants with natural light; along the southern exposure horizontal shading allows heat gain in winter but blocks it in summer. The insulation value of the curtain wall is R-8, giving four times the insulation of a conventional building.
- An exhaust system allows airborne irritants to be removed, an especially important feature in a sculpture studio.

BENJAMIN FRANKLIN ELEMENTARY SCHOOL

- Building type: Institutional/Education
- Location: Kirkland, Washington

■ Architect: Mahlum Architects
■ Completion date: 2005

About 450 students attend Benjamin Franklin Elementary School. The building is two stories, with classroom wings reaching out from two central courtyards, which provide outdoor learning environments. Teaching is carried out in small learning clusters, each having four classrooms built around a multipurpose area. Classrooms are lit by natural daylight, and skylights and large windows are operable for natural ventilation. Materials must meet old standards of durability and ease of maintenance, but great care was taken to choose low impact and nontoxic products.

Green materials
■ Rubber resilient flooring absorbs impact and is easily maintained.
■ The retro-plated concrete floor finish is nontoxic, and can be cleaned without strong detergents and chemicals. There is no need for applied floor finishes that might collect mold and decrease air quality.
■ The wool wall covering absorbs sound and has a tackable surface, which is soil and stain resistant.
■ The rooms have carbon dioxide sensors and occupancy sensors.
■ Recycled glass cullet, crushed glass that is remelted and reused, uses less energy in production and avoids waste.
■ Ground-face concrete block is long lasting, does not need to be painted, absorbs sound, and is moisture and mold resistant.
■ Paint was made from low-value volatile organic compounds, eliminating a source of pollution.
■ Walls are insulated to R-25, while the roof has an even higher rating of R-38.

Green technologies
■ Roof slopes face south and north to better control light and heat. The south facades have overhangs and sunshades.
■ Operable windows and ventilation chimneys in each classroom provide 10 air changes per hour.
■ Wool wall coverings are a biodegradable, renewable fiber product that is low in maintenance.
■ Stormwater is collected through a series of rain gardens. Rainwater is collected and held in point-source bio-retention cells.
■ The importance of rain in this area called for native, drought-tolerant plantings, which do not require irrigation.
■ Plumbing fixtures are low flush and low flow, with waterless urinals. It is estimated that the school saves 60,000 gal of water per year.

REGIONAL ANIMAL CAMPUS

■ Building type: Commercial
■ Location: Las Vegas, Nevada

- Architect: Tate Snyder Kimsey
- Completion date: 2005

The Regional Animal Campus serves as an animal shelter and adoption center for the Las Vegas area. Its goal is to display the animals in a humane, dignified manner while keeping costs low and using green strategies. Two major considerations in Nevada's climate were reducing the cooling load and water use. The dogs are housed in 22 "bungalows," consisting of 12 kennels, outdoor dog runs, and a visitors' room. Four freestanding towers with photovoltaic panels help provide cooling and ventilation.

Green materials

- Reusable components: All concrete remains of the site's abandoned water treatment plant were bulldozed, mixed, and recycled for structural backfill.
- Ferrous metals and aluminum: Less carbon promotes healthier living and better air quality.
- Radiant floor heating: Plastic tubing and forced hot water provides for thermal comfort while minimizing energy consumption.

Green technologies

- Large canopies with photovoltaic cells shade the facility and provide 25 percent of the daily power.
- Natural daylight and fresh air, necessary for the health of the animals, mandated the use of daylighting and wind-powered ventilation. Shaded windows and skylights admit a large amount of daylight and create a bright and friendly atmosphere, while lowering electrical cost.
- Reclamation system: A water treatment plant treats all wastewater for reuse on-site. It recycles and filters water, reducing by 50 percent the amount of water needed for the center. Twenty two thousand gallons of water is recycled daily through the on-site water reclamation system.
- Operable exterior louvers open automatically, allowing fresh air to circulate through the building.
- The wind-powered ventilation system works by forcing cool air in at a lower level of the cooling chimneys/towers. As the heat rises to the top, it is pushed out through the vents. The tall chimneys also serve as signage. This system significantly reduces building cooling loads while providing 100 percent fresh air.
- Radiant floor heating and efficient HVAC units produce interior comfort.

PHILIP MERRILL ENVIRONMENTAL CENTER

- Building type: Commercial/Education
- Location: Annapolis, Maryland
- Architect: Janet Harrison, Greg Mella (SmithGroup)
- Completion date: 2000

The Philip Merrill Environmental Center is a 32,000 ft^2, two-story building with a wood frame, situated by Chesapeake Bay. Its mission is to "protect and restore the Chesapeake Bay Area." The building planners expect the facility to have little or no impact on the surrounding areas. Materials used in the building are either recycled or produced through methods that are not harmful to the environment.

Green materials

- Recycled galvanized roofing material is recycled content made from aluminum. The roof holds an R-value of 25.
- Recycled-wood fiberboard and particleboard can be used as a sound barrier between walls and floors.
- Natural linoleum flooring is long lasting, self-curing, renewable, and biodegradable.
- Cork sheets under the linoleum adhere together and act as insulation beneath the flooring.
- Bamboo flooring is used in the main foyer.
- VOC-free paint, containing no volatile organic compounds, was used throughout.
- Planted swales are used instead of curbs and gutters.
- Structurally insulated panels (SIPs) with insulating foam are used in place of studs and rafters.

Green technologies

- Galvanized siding was made from recycled cans, cars, etc. It is rust free, due to a hot dip process that when dry protects the ferrous metal.
- Natural daylighting is achieved by large windows, clerestories, high ceilings, and an open interior, allowing daylight to penetrate the entire building.
- Geothermal wells (48,300-ft-deep wells) use the earth's constant temperature to regulate the indoor climate. It uses the earth in the summer as a heat sink and in the winter as a heat source.
- Photovoltaic panels, located on the south wall, produce energy on-site.
- Louvers, awnings, or trellises provide shade, reducing the solar heat load.
- Low-water-use fixtures use automatic faucet controls that sense movement to operate.
- Zone HVAC systems allow heating of only specific areas, rather than the entire building.
- Roof solar panels provide all domestic hot water for the building. This saves 120 kWh of electricity a day.
- Incentives for bicycle use as opposed to vehicle use are seen in the bicycle racks and bicycle storage, showers, and changing rooms. There is free battery charging for those with electric cars.
- Parking space was moved underground and existing pavement was demolished and replanted with native plants.
- A rainwater collection system collects water that is used in bathroom sinks and fire suppression.

ALBERICI CORPORATE HEADQUARTERS

- Building type: Commercial
- Location: St. Louis, Missouri
- Architect: Mackey Mitchell Associates
- Completion date: 2004

The Alberici Corporate Headquarters building is a conversion of an old manufacturing facility, which followed "deep green" practices to create a two-story, 110,000 ft^2 office space in a mezzanine style. Open space design was the goal, and interiors are clustered around three large atria, which encourage better ventilation.

Green materials

- Native plants used on the site are low maintenance and eliminate irrigation. The use of local earth resources is both ecological and sustainable.
- Brick is used for much of the building, as its high durability, low cost, and efficient construction make it very sustainable.
- Operable clerestory glazed glass windows allow for natural ventilation and daylighting.
- The steel frame is a recyclable material as well as strong and resilient, making it a durable material.
- Bamboo ply-board, particleboard made from wheat, and flooring of cork are rapidly renewable resources.
- Recycled-content materials, such as fly ash, steel framing and rebar, rubber flooring, and carpet backing were used.

Green technologies

- An on-site wind turbine and solar water-heating system help to power 20 percent of the building. They both incorporate high performance technologies.
- Water-efficient fixtures reduce consumption by 70 percent, saving up to 500,000 gallons of water a year.
- A cooling tower and sewage conveyance system uses runoff rainwater from the roof to operate and cool the building. This well-designed system is ecological because of its low pollution and emissions, as well as its use of renewable earth resources.
- Fifty seven percent of the materials needed in the project were local to the site.
- Skylights, three shades of glass, and attention to how windows were placed give 90 percent of the occupants outdoor views.
- Under the floor an air-distribution system allows occupants to receive air at their level.

BALLARD LIBRARY AND NEIGHBORHOOD CENTER

- Building type: Public
- Location: Seattle, Washington
- Architect: Bohlin Cywinski Jackson
- Completion date: 2005

This branch of the Seattle Public Library consists of a 15,000 ft^2 facility with accompanying neighborhood center and underground parking. An outstanding feature of the building is its broad, planted roof, which opens clerestories underneath, and floods the interior with natural light. It focuses on the young, diverse population of the community, while engaging and encouraging human interaction. The main objective of this project is to demonstrate green building on a small budget, thus raising awareness of environmental issues.

Green materials

■ Materials with an integral finish, such as stainless steel is used, requiring less assembly time and waste. The concrete flooring only requires a seal finish.

■ Solar panels assist in shading the lobby as well as produce energy for lighting and reducing CO_2 emissions. These solar panels wrap around the floor-to-ceiling windows in the lobby that receives the most sun exposure. Light still penetrates through portions of the glazing, saving energy in lighting loads, while still shading the room and reducing heat gain.

■ Concrete is mixed with recycled milk cartons to make the concrete more durable and less polluting to the environment by reducing the amount of cement used.

Green technologies

■ Occupancy sensors activate the artificial lighting only when the space is being used. This sustains the life of the lighting system and saves energy.

■ Steel tubes inside the library support a laminated-beam ceiling, with suspended ducts for ventilation. The complete open space allows a "library without borders."

■ The structure is of steel with high durability factors, while the exterior cladding is a combination of metal/glass curtain wall.

■ Waterless urinals, along with low-water-use fixtures and automatic faucet controls are used throughout the building.

■ Low-water-use fixtures save water due to limiting waste.

■ Automatic controls for the irrigation system detect when water is needed and only then activate the irrigation system. This sustains the life of the whole system due to less usage and saves water because the plants never get excessive watering.

■ The garden roof is landscaped with local plants that are also drought resistant. The vegetation enables 86 percent of the storm runoff to be absorbed, as well as reducing heat gain.

BANK OF AMERICA BUILDING

■ Building type: Commercial
■ Location: New York City, New York
■ Architect: Cook + Fox
■ Lead designer: Serge Appel
■ Completion date: 2009

Figure 8.13   Ballard Library : Two external and one internal views of Ballard Library.

The Bank of America building houses the New York City headquarters of the Bank of America and additional leased office space. It is the second tallest building in the city, with an architectural spire of 255.5 ft. Cook + Fox Architects not only considered the occupants when designing this building, but also the way people use the outside surrounding area. The building improves the pedestrian and transit circulation with widened sidewalks, through-block passageways, and an urban garden.

Green materials

- Floor-to-ceiling, translucent, high-performance glass ensures maximum daylighting in interior spaces. This promotes a healthier work environment, both physically and mentally.
- The crystalline facade minimizes obstruction of views and does not block sunlight for the surrounding buildings and street, while capturing and refracting the changing angles of the sun.
- Recycled and low VOC materials are used to reduce off-gassing.
- A mixture of slag in the concrete lowers the amount of CO_2 that traditional cement manufacturing produces.
- LED lighting is used throughout the building to ensure energy efficiency.
- Waterless urinals and low-flow fixtures were installed to conserve water.
- A double wall used on part of the facade helps to conserve energy.

Green technologies

- Glass acts as an insulator, minimizing solar heat gain and loss.
- A thermal storage system works like a "battery" for cooling. In the basement, there are forty-four 10-ft high, cylindrical tanks with water and a cooling coil inside. At night the water freezes, and during the day the ice melts, cooling the building.
- Rain and wastewater is captured and reused to fulfill nearly all of the needs of the structure, saving 10.3 million gallon annually.
- An under-floor air delivery system, which efficiently provides for natural ventilation and highly filtered fresh air, also facilitates a more controllable, healthful, and efficient heating and cooling system.
- An on-site cogeneration plant works in conjunction with the thermal storage system. At night, the electrical production from the cogeneration plant exceeds the building's needs, and the excess energy is used to run the chilling equipment to freeze the water in the cylindrical tanks. This shifts the electrical load from daytime to nighttime, reducing the impact on New York's electrical grid.

BEDDINGTON ZERO ENERGY DEVELOPMENT (BedZED)

- Building type: Commercial
- Location: Beddington, Great Britain
- Architect: Peabody Trust, Bioregional Development Group, Bill Dunster Architects
- Completion date: 2002

BedZED is a net zero housing development. It is promoted as "Britain's largest carbon-neutral ecologically friendly community." The goal behind this project was to create buildings that would produce as much energy from renewable sources, such as recycled wood and solar power, as it consumes. The development comprised of 82 homes, as well as offices and many other mixed-used spaces. The layout of the development takes into account the sun, solar gain, and light by facing all the homes to the south. Each home features large glass openings to let in natural daylight and maximize solar gain, as well as a terrace with green space.

Green materials

- Wood chips are used for the CHP heating system.
- FSC certified wood is used for building material and panels. FSC uses forestry that is practiced in an environmentally responsible way that takes into account economic viability and social responsibility.
- Roof gardens and garden terraces collect rainwater and provide ambient cooling.
- Low-E glass reflects a significant amount of radiant heat, which lowers the total heat flow through the windows.

Green technologies

- CHP: The combined heat and power plant is an energy cogeneration power generator. The unit gets the heat energy generated by unwanted wood chip waste, which is converted to electricity. The heat energy is captured in hot water and distributed through a system of superinsulated pipes.
- All portions of the building that are not glazed are wrapped in 30 cm of insulation.
- A reed-water biofiltration system purifies blackwater into gray water, which is then used for nonpotable use.
- A photovoltaic array of cells converts solar energy into direct-current electricity, which reduces the consumption.
- Heat sinks are used in a wide range of applications, such as, heat engines, refrigeration, and cooling electronic devices.

BIP COMPUTER BUILDING

- Building type: Commercial
- Location: Santiago, Chile
- Architect: Alberto Mozo
- Completion date: 2007

In designing the BIP Computer Building, Architect Albert Mozo wanted to make the structure as moveable and reuseable as possible, while not disturbing the existing buildings on the site. His goal was to make sure "the building would be easy to take apart." The reason for this unusual goal was that a larger building will probably be erected in the future. Eighty percent of the two neighboring buildings were retained and renovated to be used as storage and customer service. An interesting feature of the new structure is that rather than being anchored, it rests on a concrete platform which gives it a feeling of "floating" one and a half feet above the ground. The design also embraces the "transitivity" concept, meaning it has the ability to change over time.

Green materials

- Laminated pine timbers, all of the same size, which were then bolted together, comprise the building's structure. These are ordered precut from the factory and can be disassembled and reused in a new building. Much of the wood came from renewable sources in Chile.
- Laminated glass with a middle layer of napa, a fiber used in bed covers and jackets, makes the glass a translucent white, limiting glare and heat brought into the work stations.
- Precast concrete pavers on all three floors slow changes of temperatures in rooms with their thermal mass.
- Concrete material was made of a conglomerate of rocks, gravel, and sand, providing a strong building material.

Green technologies

- The same wooden beams that comprise the structure are also used for the ceiling beams. The walls are completely of glass.
- The concrete flooring tiles can be easily recycled.
- All interior divisions were eliminated for maximum flexibility. This allows the various functions of the space to manipulate the interior, and limits the amount of materials used.
- Unskilled laborers could do much of the labor owing to the standardized and simplified construction methods.

"ENERGY TOWER" BURJ AL-TAQA

- Building type: Commercial
- Location: Dubai, United Arab Emirate
- Architect: Eckhard Gerber
- Completion date: 2009

German architect Eckhard Gerber designed "a giant 68-story building projected to rise 1056 ft, which would make it number 22 on the list of the world's tallest buildings." But more importantly, the overall idea for this building was to create a commercial high-rise that produces no emissions and can runoff its own energy source. The "Energy Tower" design will contain a 200-ft wind turbine and an array of photovoltaic panels. The environmentally friendly building will create a place for many people to work under a sustainable design. With all of the green technologies and materials used in this project, it is expected to use a great deal less energy than traditional buildings of its size.

Green materials

- Vacuum glazed glass prevents glass from absorbing heat from the extreme temperatures that can reach 122°F.
- Seawater will be used to cool the building by running it through a series of pipes, allowing the temperatures inside to be lowered to 64.4°F.
- Photovoltaic panels are located on the external glass walls, providing the building with a considerable amount of energy.
- Steel building material will provide the building with sustainability long into the future.
- Using concrete as a foundation means the "Energy Tower" will be very durable and will support the structural integrity.
- A natural air-conditioning system uses the Iranian wind to cool the interior spaces of the building.

Green technologies

- The vertically mounted wind turbine on the top of the high-rise creates a substantial amount of energy. It pulls air down through the internal spaces, making a more comfortable environment.
- The rotating solar shield is designed to track the sun's rays and collect energy through its semitransparent photovoltaic. This system contributes to the building's green structure and gives the building a maximum amount of solar power with the least amount of materials.
- Hanging gardens called atria help clean the air.
- The double-layered façade allows wind to travel throughout the building, reducing energy use by 40 percent in comparison to other conventional high-rises.
- The cylindrical shape was chosen to expose less of the facade to the glaring heat of the country.
- A duct system, along with openings on the facade of the building, aid in ventilation by pulling in fresh air.
- Solar panels, drifting in the sea near the building, provide additional energy.
- Mirrors on the roof direct natural light onto a cone of light that is directed through the central part of the building.

C.K. CHOI BUILDING

- Building type: Commercial/Educational
- Location: Vancouver, British Columbia
- Architect: Kubala Washatko
- Completion date: 1996

The C.K. Choi Building for the Institute of Asian Research houses five research centers that are occupied by approximately 300 people. With a concentration on recycling materials and attempting to reduce water usage as much as possible, the C.K. Choi Building is one of the successful advanced green buildings.

Green materials

- Salvaging and reusing heavy timbers from an adjacent demolished building for 90 percent of the Choi structure promised durability and recycling benefits.
- Sinks, doors, toilet accessories, and some electrical conduits were also reused from the previous building.
- The majority of nonstructural steel and all the bricks used in construction were rescued and recycled.
- Double-glazed windows, as well as insulation on the exterior, save on thermal gain.

Green technologies

- Composting toilets, which dispose of the solid wastes by using organic matter such as wood chips, save 264 gal of water each day. Wastes end up as a humus soil that enriches the earth.
- All the water reductions made it possible to build without connecting to the main sewer system.
- Rain is collected in an 8000-gal cistern and then used for irrigation.
- Daylight and occupancy sensors control the consumption of power used to light the building.
- An adjacent building, which produces excess power, handles the power needs of the Choi building. No new electrical service or transformers were needed.
- Waste heat, collected from the campus underground steam infrastructure, is used to heat water for the building.
- Treatment of wastewater is integrated into the landscape design. Gray water enters a trench with marsh vegetation.
- Walkways are interlocking pavers on a gravel base, which aid in promoting surface runoff.
- Gingkoes, which clean the air of pollutants, were chosen for landscaping. Existing trees were integrated into the new landscaping.
- Orientation of the structure was adapted to the prevailing winds and the peak hours of sun.

CALIFORNIA ACADEMY OF SCIENCES TRANSITION FACILITY

- Building type: Commercial/Public
- Location: San Francisco, California
- Architect: Melander Architects
- Completion date: 2006

The California Academy of Sciences Transition Facility was constructed as a temporary building for the purpose of research, administrative, and public space while the new facility is being built. The first floor is used for the aquariums, the second contains exhibits and an interactive educational center, while the third floor is for administrative use. The fourth through sixth floors house collections. The temporary space, which will be used for five years, is designed to have little external impact on the environment. Melander Architects planned with the goals of reuse, economy, and efficiency in mind.

Green materials

- Recycled theatrical lighting.
- New compact storage systems were chosen with regard to the needs of the future building.
- Plexiglas, a strong, break-resistant material, was reused.
- The transition facility used steel to support the overall structure in hopes that the building could continue to be used by a new consumer when the move into the new facility takes place.
- Perforated birch was used in the building in its raw form. It did not get an overcoat compound, which can release toxic vapors into the environment.
- Plastic laminate was selected as inexpensive, waterproof, attractive, and relatively easy to install.

Green technologies

- Adaptive reuse: Many items were reused, but using items from the old facility, changing it for the temporary facility, and then manipulating it again for the new facility made the transition economical.
- Concrete walls were left bare since the building will be used for such a short time, which cut costs substantially.
- Many other components in the building were left in their raw form to alleviate costs. However, the practice also promotes a nontoxic interior.

LONDON CITY HALL

- Building type: Municipal/Public
- Location: London, England
- Architect: Foster and Partners
- Completion date: 2002

London City Hall is the headquarters for the Mayor of the London Assembly. The building is constructed of rounded glass, presenting a truly stunning sight on the edge of the Thames River. Besides governmental use, the building houses offices, shops, and an amphitheatre. On the inside a circular stairway takes one to the extreme top of the building, where a balcony gives a superb 360° view of London. The unique shape was planned to reduce the overall area and thus improve energy efficiency.

Green materials

- Glass panels of triple-glazed low-E glass were laser cut.
- The interior is flexible; occupants are able to enjoy an open floor plan or to convert the space to cellular offices.

Green technologies

- Air-conditioning comes from ground water pumped from the water table through bore holes. The same water is used in restrooms and for irrigation purposes.
- The form of the building is angled to reduce both solar gain and heat loss.
- Photovoltaic solar panels provide up to 70 percent of the energy. They are expected to reduce CO_2 emissions as well, perhaps up to 3000 tons during the building's lifetime.
- Natural ventilation cools the entire building.
- The distinctive rounded and "stepped" facade provides shade.
- The building uses only 25 percent of the energy a conventional building of this size would require.

MONT-CENIS ACADEMY

- Building type: Educational/Public
- Location: Herne Sodingen, Germany
- Architects: Jourda, Perraudin, Hegger & Hegger Schleif
- Completion date: 1999

Mont-Cenis Academy is a training academy that also contains seminar facilities, meeting rooms, a restaurant, a gymnasium, a library, and leisure facilities, all under a glass shell. As the winning design at the design competition in 1991, it took 10 years to complete. It is a beautiful transparent building with a front overhanging canopy.

Green materials

- Photovoltaic modules and glass panes are glued onto aluminum profiles. They provide electricity by collecting solar energy and converting it to electricity.
- The photovoltaic cells in the glazing of roof and façade also act as a sunscreen for the building and create a microclimate envelope for this "glass house." They also provide shade for the interior.
- The wood framing consumes less total energy in its lifetime than buildings made of steel or concrete.

Green technologies

- Photovoltaic modules control the interior climate of the glass house by creating a microclimate glass envelope, controlled by different cell densities in the modules of the overhanging glazing.
- Installed to reduce energy consumption in winter and summer it is an air handling system unit with a heat exchange system.
- In summer, the roof and façade can open, providing ventilation and reducing the need for air-conditioning.
- Seven ground channels help to bring in ventilation from the outside by collecting breezes and directing them into the building.
- The glass and photovoltaic-module envelope provides a garden-like interior with a mild microclimate.
- About 600,000 kWh of measured energy is the annual output.

NORDDEUTSCHE LANDESBANK AM FRIEDRICHSWALL (NORD/LB)

- Building type: Commercial
- Location: Hannover, Germany
- Architect: Behnisch, Behnisch & Partner
- Completion date: 2002

The design for this high-rise bank building was the winner in an international competition for the project. Several aspects garnered it the prize: the use of daylight inundating the building; its care for the comfort of the workers; and very importantly, its energy efficiency. Two goals for efficiency were set forth: to better the regulations for insulation by 10 percent and to "create environmentally sensitive measures at reasonable cost."

Green materials

- The roof garden includes herbaceous plants and roses, and helps collect rainwater.
- Recycled steel drastically reduces embodied energy.
- The ventilation system is nearly completely natural. Shutters from the courtyard supply fresh air.
- Low-E glass reduces the U-factor by suppressing radiative heat flow.
- The steel/glass curtain wall as exterior cladding lets natural lighting in the building, which helps reduce energy use and is durable as well.
- A radiant slab cooling system and a geothermal heat exchange is used.

Green technologies

- The courtyard's large area of water and the green roof of the cafeteria supply fresh air from its microclimate that invigorates the building's interior.
- Air is vented into office spaces and then to chimneys or vents that exhaust air to the roof.
- The superstructure is exposed and the windows provide ventilation. Double façade areas provide protection against noise and vehicle emissions, while serving as a duct transferring clean air from the central courtyard.

PEARL RIVER TOWER

- Building type: Commercial
- Location: Guangzhou, China
- Architect: Skidmore, Owings & Merrill (SOM)
- Completion date: 2009

Pearl River Tower was designed to be "as sustainable as possible," a net-zero-energy skyscraper with virtually no environmental impact. Though the client was a tobacco company, SOM Architects saw no reason why the 71-story skyscraper could not have a green design. "What we have is a series of small steps that get you to something that makes a difference." To do this, the firm focused on four goals: reduction, reclamation, absorption, and generation.

In the reduction phase low-E-glass was incorporated on the southern facade with double-layered curtain walls, and the tower was rotated to the east. In the reclamation phase the designers looked at how the building could use its own energy to do double duty; for instance, the chilled slab concrete ceilings enhance daylight but also cool the air. The main absorption technique uses a geothermal heat sink so that water could be naturally cooled with little energy spent. This naturally cooled water could then be used in the cooling towers. As the designers looked at generation of energy, they decided on wind power, photovoltaics, and microturbines.

Green materials

- Low-E glass (having a low emissivity) prohibits extreme temperature fluctuations that regular glass transmits.
- Double-layer curtain walls limit the amount of direct light into the interior, controlling the heat in the building.
- Photovoltaic panels, which create energy from sunrays, were placed on the building's façade situated toward the most sunlight.
- Chilled slab concrete vaulted ceilings regulate temperature in the building by cooling the air.
- Radiant panels, used with a low-energy, high-efficiency system, redirect light to different areas limiting electrical light use.
- Operable skylights let in light and fresh air.

Green technologies

- On the tower there are several different exterior envelopes, including a photovoltaic system and the double-layer curtain wall.
- Exhaust air from each floor is routed into the double-layer curtain wall cavity.
- Gray water retention allows for used water to be recycled, limiting water use.
- The high-efficiency lighting system uses radiant panel geometry to distribute light to different areas.
- Air is forced through wind turbines which speed up the wind velocity, thus producing 15 times more electricity.
- Automated blinds maximize or minimize the light let into the building.

NATIONAL AQUATICS CENTER (WATER CUBE)

- Building type: Commercial
- Location: Beijing, China
- Architect: PTW Architects and Arup
- Completion date: 2007

The Water Cube, or National Aquatics Center, was built in Beijing for the 2008 Summer Olympics. But the building provided public multifunction activities before the games began and it continues to be used. The elements of water and bubbles provided inspiration for the lightweight design, which looks like "a rectangular box covered in iridescent bubble wrap." When receiving an award at the International Architecture Exhibition, the building was recognized as, "a stunning way of morphing molecular science, architecture, and phenomenology to create an airy and misty atmosphere for a personal experience of water leisure."

Green materials
- ETFE is the translucent plastic that makes up the bubble cladding. It can hold up to 300 times its weight without trussing. It is able to span greater distances than glass, costs 70 percent less to install, and is recyclable.
- The ETFE bubbles are easier to clean than glass.
- The durable steel frame, clad with ETFE pillows, allows more heat and light penetration than glass, making heating the building more efficient.

Green technologies
- The choice of ETFE bubbles has resulted in a 30 percent reduction in energy costs. This is equivalent to covering the entire roof in solar panels.
- A rainwater collection system gathers 10,000 m^3 of water a year, while the recycling system reuses 80 percent of the building's water. All the backwash water is filtered and returned to the pool.
- LED-lit bubbles act as adjustable insulators, turning the building into a greenhouse. They serve as storehouses for warm air that can be pumped into the Cube as needed.

ALDO LEOPOLD FOUNDATION HEADQUARTERS/ LEGACY CENTER

- Building type: Commercial
- Location: Fairfield, Wisconsin
- Architect: Kubala-Washatko Architects
- Completion date: 2007

The Aldo Leopold Center was established to honor Leopold, the author and environmentalist. The Foundation embodied Leopold's "Land Ethic" principles in the design and function of the Center. Besides serving as headquarters, it provides tours, workshops, and research programs. It is also the first LEED-platinum carbon neutral building, and, arguably, "the greenest building ever built, with zero footprint and great design."

Green technologies
- A 198-panel 39.6-kW solar electric system produces 15 percent more energy than what the building requires.
- The concrete floors are installed with a radiant system which both heats and cools the building.
- Geothermal energy, excellent insulation, and a design that permits natural lighting and shading aid the radiant floor heating system in keeping temperatures stable.
- Cross ventilation and operable windows efficiently cut down on cooling costs.

Green materials

■ Structural insulated panels provide a well-insulated building envelope that is highly efficient.

■ Glass used throughout the center was chosen for its high R-values, making insulation properties throughout more than double local code requirements.

■ Much of the wood that was needed for construction was harvested locally, diminishing the embodied energy. The wood that was harvested on-site was sent to a local mill.

■ Stone was reused from a previous building on-site, and limestone that was originally used in the construction of a nearby airport hanger is now a fireplace and rainwater aqueduct.

■ Seventy eight percent of the wood is FSC certified.

■ Recycled-content cellulose insulation was used.

■ The aluminum roof has a durability unsurpassed by other options.

■ Interior walls used a plaster made with clay, straw, and local sand.

CESAR CHAVEZ LIBRARY

■ Building type: Educational/Public
■ Location: Laveen, Arizona
■ Architect: Line & Space, LLC.
■ Completion date: 2007

The Cesar Chavez Library, a tribute to the labor activist and farm worker Cesar Chavez, is an advanced green building designed by Line & Space, LLC for the town of Laveen, to serve as a modern public library. The building has a winglike roof and curved walls, which simulate the curve of the nearby lake. It makes a striking contemporary statement. The building program consists of one main floor and two entrances, including a children's area, a teen lounge, and a 75-seat meeting space for the public.

Green materials

■ Locally available recyclable concrete, steel, and aluminum provide low-maintenance materials. Energy consumption in transportation is low.

■ The construction materials were chosen because of their natural and durable qualities.

■ Landscaping uses natural plants in the desert, low-water area to decrease water consumption.

■ Roof collection systems and channels bring water from the roof directly to trees.

■ A gray water collection system is used to reuse water and decrease overall water usage.

■ Earth beams result in greater structural thermal mass and decrease the weight load of the building.

Green technologies

- Passive solar design prevents heat gain through the use of sunshades and structural overhangs.
- Conventional HVAC systems are affordable and protect users from the harsh Arizona desert.
- Location between two earth beams on the east and west helps regulate the hot desert heat.
- The south/north orientation of the building also helps organize the daylighting in the building, requiring less electric lighting, while large roof overhangs prevent overheating and solar glare.
- A rainwater collection system is employed, which collects rain from the roof and stores it in the adjacent reservoir. A water pump irrigates the landscaping.
- Waterless urinals and low-flow toilets reduce the need for excess amounts of water.
- Water use is dramatically reduced by a number of low-flow and efficient appliances.
- Faucets in main facilities are automatic and sensory in order to reduce water waste.
- The construction process included a waste reduction plan.
- Because the library is located near two major schools and public transportation, the need for car usage is reduced.
- Hot water is provided only to staff members, thus conserving energy.
- Electrical systems are equipped with economizers in order to enhance efficiency by gauging the daily temperature needs.

CHICAGO CENTER FOR GREEN TECHNOLOGY

- Building type: Commercial
- Location: Chicago, Illinois
- Architect: Farr Associates Architecture
- Completion date: 2003

The Chicago Center for Green Technology (CCGT) is a green building with a high level of performance. The Center helps people of the community, including professionals and homeowners, learn about the cost-effectiveness of green technology and its many benefits for environment and people. The building is a two-story L-shaped structure with a program that includes various spaces for public seminars, which are offered two or three times a week, and a small library. The occupants supply environmental products as well as services. Green-corps (the city's community gardening and job training program) and WRD Environmental (an urban landscape company) both have offices at Chicago Center. The CCGT uses 40 percent less energy than a similar building of the same size, saving approximately $29,000 a year.

Green materials

- Green features of the building include sunscreens on the south side with exterior louvers, awnings, and trellises.
- The interior lights have sensors that gauge the amount of natural light coming in from the outdoors and dim accordingly; as the sun goes down the lights slowly grow brighter.

- Natural daylighting from large, specially glazed windows and skylights reduce the use of artificial light by 24 percent.
- The existing infrastructure of the previous building was able to be reused, creating less waste and using less material.
- Recycled glass tiles, rubber wall base, reclaimed lumber, and recycled plastic toilet partitions were low-impact choices.

Green technologies

- The majority of materials came from within a 300-mile radius, helping reduce harmful emissions from transportation.
- A rooftop garden and a natural habitat collect and filter stormwater.
- Geothermal energy and solar energy in the form of roof panels are combined to power the building.
- Over 40 percent of the materials used for the construction came from recycled materials.
- The indoor air quality is monitored very closely, and the duct size has been increased.
- Ground coupled systems include heat pumps as a source of heating and cooling.
- Sensors control the HVAC, ensuring excess energy is not used.
- The building reuses rainwater and other groundwater.
- Solar energy provides 20 percent of the building's energy needs.

EDEN PROJECT

- Building type: Commercial
- Location: Cornwall, Great Britain
- Architect: Nicholas Grimshaw
- Completion date: 2001

The Eden Project in Cornwall, England, is well named, as it is a multidomed greenhouse that supports plant species from all over the world. Each dome is a duplication of a natural ecosystem; for instance, one dome simulates a tropical environment. The shape of the domes, which was inspired by the moon, are numerous hexagons connecting the entire structure. Each hexagon forms a transparent cushion made of plastic.

As Director Tony Kendle states, "Eden isn't a destination; it's a place in the heart. It's a statement of our passionate belief in an optimistic future for mankind." The green approach that the Eden project envisions is to be a model for ecofriendly interaction.

Green materials

- Some materials from landfills were used to create sculptures around the site.
- ETFE were used because they have a high corrosion resistance and strength over a wide range of temperatures. ETFE also has a very high melting temperature and, unlike other plastics, does not emit toxic fumes when ignited. This plastic also transmits more UV light rays into the structure.

- In the core building the roof was made from FSC certified red spruce. Photovoltaic panels on the roof contribute to green energy. The copper covering the roof comes from a mine that has one of the highest environmental and social standards of any in the world.
- The use of local rocks for biomorphic sculpture is also a green approach, and reduces the emissions of CO_2 into the atmosphere.
- The core building uses Glulam, a glue-laminated lumber which is made from veneer bonded together with an adhesive. Glulam is incredibly versatile and is one of the strongest structural materials per unit of weight. It also generates no waste because its off-cuts are used as a fuel.
- The frame is made of steel and thermoplastic.

Green technologies
- Energy from a local wind farm supplies the power.
- The car fleet runs on cleaner burning propane, which does less harm to the environment.
- At the Eden project local vendors provide supplies, helping to invigorate the local economy and to reduce transportation energy and emissions of CO_2.
- Rainwater is collected and used to flush the toilets.
- Automatic taps are installed in sinks to save water.
- On the roof, limestone filters the water runoff to remove any copper taint. This helps to clean the water so that it can be used in the building.
- A lobby reduces heat loss from the building.

EDIFICIO MALECON

- Building type: Commercial
- Location: Buenos Aires, Argentina
- Architect: HOK Sustainable Design
- Completion date: 1999

The Edifico Malecon is a glass office tower that was designed to be "the cutting edge of key technologies." The building is long and narrow, a slab that minimizes surface solar gain, which is important in Argentina's climate. The slab is curved or pinched in at the ends to provide better all-around views. The building is primarily composed of steel and concrete, which, though not particularly ecological, is very sustainable.

Green materials
- The materials used are glass, reinforced concrete, aluminium, and ceilings from indigenous wood.
- Local materials are used.
- A green roof with native plants cools the building and cleans the water.

Green technologies
- A high-performance curtain wall on the south side reduces solar penetration.
- An extremely efficient HVAC system uses less energy, as it is more efficient, and utilizes natural gas, a cleaner fuel than coal.

- Sunshades reduce the amount of sun penetrating the interior and the amount of cooling needed.
- Operable windows can be controlled by individual workers rather than having one uniform setting for the whole building.
- It was decided to utilize the preexisting concrete basement as a lower-level parking garage.
- The building uses the winds coming in from the river to ventilate offices and stairwells through many operable windows.

ARTISTS FOR HUMANITY EPICENTER

- Building type: Assembly/Educational
- Location: Boston, Massachusetts
- Architect: Arrowstreet, Inc.
- Completion date: 2004

The AFH EpiCenter is an advanced green building designed by Arrowstreet for the nonprofit organization Artists for Humanity. The building is 4-story high; the main floor is used as a gallery space and is often rented out for functions, while the upper floors serve as artist studios for the underprivileged teens that are working with Artists for Humanity. The organization's mission statement is "to bridge economic, racial, and social divisions by providing underserved youth with the keys to self-sufficiency through paid employment in the arts." Artists for Humanity wanted a building that exemplified their mission but also to design a building with a lasting, but low-carbon footprint. Materials were chosen based on their sustainable, ecological, and performance properties.

Green materials
- Glass is extensively used for its durability, affordability, and low maintenance. Glazing protects the glass from external conditions.
- The recyclable concrete can be reused after the life of the building and presently provides good insulation.
- Recyclable steel is also affordable, durable, with low maintenance, and can be utilized in a later construction.
- Salvaged railroad tracks are used for exterior support railings and salvaged police car windshields are used for interior railings, which not only minimizes the amount of landfill material but is also a creative and effective way to give normally useless products another life.
- Eighty percent of the total construction debris was reused in the construction of the interior. A local artist was commissioned to supervise this procedure.
- Low VOC paints and sealants are used, along with materials with little or no need for finishing applications, in order to decrease the level of toxicity within the building.
- Photovoltaic panels provide the roof with shade and help minimize the total electric bill of the building.

Green technologies

■ Passive design in the building's orientation includes south-facing windows for Boston's cold winter temperatures.

■ Windows are operable, enabling manual cooling and reducing some energy use as well as the need for extensive cooling systems.

■ The building's envelope is very highly insulated to block heat and retain coolness.

■ Daylighting and the use of clear panels as partitions allow sunlight to reach far into the building and reduce the need for artificial lighting.

■ A roof collection irrigation system waters the small square of front lawn at the front of the building.

■ There are no mechanically operating HVAC systems, which dramatically reduces the building's carbon footprint and allows the building to profit from the use of natural processes, such as manually opening the windows at night during the summer.

■ The building enjoys an 80 percent reduced energy use due to photovoltaic panels and highly efficient electrical appliances and systems, such as heat water pumps and a heating recovery unit, which takes advantage of air exhaust while using fresh air to cool or heat. Excess energy is sold to the grid in the summer.

■ An automated light control system provides light only as needed.

GARTHWAITE CENTER FOR SCIENCE & ART

■ Building type: Commercial
■ Location: Weston, Massachusetts
■ Architect: Ellen Watts, Dan Arons; Architerra, Inc.
■ Completion date: 2007

The Garthwaite Center for Science and Art has a varied program, including classrooms, offices, meeting spaces, a science display atrium, and the campus data center. It was designed to fit in a small area while preserving as many trees as possible. The levels of each floor are arranged to conform to the slope of the hillside. The Garthwaite Center for Science and Art was chosen as an AIA Committee on the Environment Top Ten Green Projects in 2008.

Green materials

■ The materials used were primarily from local places, reducing transportation costs.

■ The partial green roof is successful in that it provides an economical way to help heat and cool the building. The overhangs on the roof help protect against heat gain.

■ The use of glass on the exterior is an ecologically sound choice to reduce the use of artificial lighting.

■ The photovoltaic panels provide clean energy.

■ Recyclable materials were used for glass tile and countertops.

■ Recycled wood-fiber wall panels saved the material from waste sites.

■ The exterior wood was treated with a heavy-bodied stain for durability. The timber now has a honey-colored look, but it is a personal choice as to whether to maintain it; it would still function well without maintenance.

■ To conserve water, Xeriscape Landscape and waterless fixtures like composting toilets were selected.

Green technologies

■ The use of drought-resistant meadow mix plants and native plants need no irrigation and make an ecological contribution in reducing water usage and helping to clean the air.

■ The use of T5 fluorescent lighting and light-colored surfaces were used to reduce the electric lighting load.

■ The building is oriented to the south to take advantage of passive solar heating.

■ Flexible space is achieved through hinged wall panels that are movable, creating a wide use of space within the building.

■ Radiant floor heating is efficient and provides a huge energy savings performance. A boiler that burns wood pellets provides a 67 percent reduction in green house gases when compared to conventional boilers.

■ Since the building maintains its internal temperature far longer than conventional buildings, it is the designated shelter for campus-wide emergencies.

■ The architects left some of the building's systems exposed for educational purposes. The enthalpy heat wheel, wood pellet boiler, and toilets composters can be seen by students.

■ Stormwater is managed by the use of subsurface infiltration basins.

HEARST TOWER

■ Building type: Commercial
■ Location: New York City, New York
■ Architect: Foster + Partners
■ Completion date: 2006

The Hearst Tower, headquarters for the Hearst Corporation, has been called the "greenest skyscraper in New York City," soaring 44 stories high in a distinctive triangle-shaped framing. In 2008, it was awarded the International Highrise Award. The multifaceted exterior is impressive and makes possible an extremely open interior, described as a "vast internal plaza," with 360° views of the skyline. Foster + Partners states that the diamond design of the steel and glass exoskeleton uses less steel than a traditional grid skyscraper. Two major challenges of the designers were to meet goals of urban infill and adaptive reuse. The Hearst Tower's strengths are in the performance and sustainability aspects of the building.

Green materials

■ Most of the steel used was recycled.
■ The majority of the materials were locally supplied.

- Almost all the waste from construction was recycled.
- Ninety percent of the structural steel contains recycled materials.
- The atrium contains a three-story water sculpture in the shape of a waterfall, which both cools and humidifies the area, using rainwater collected at the rooftop.
- A multiplicity of energy-efficient mechanical controls were installed, along with "the integration of every sustainable feature available" in 2006.
- The triangular shapes required only about 80 percent as much steel as conventional high-rises.
- A fully automated lighting system was installed.
- Energy-efficient appliances more than meet standards set.

Green technologies

- Polyethylene tubing under the floor circulates water for the purpose of cooling and heating the building.
- The atrium floor is heated limestone.
- The tower uses 26 percent less energy than the minimum energy use regulated by New York City.
- A tank in the basement collects rain from the roof which is used for the cooling system, irrigation, and the fountain in the lobby.
- The building can be ventilated naturally for three-fourths of the year.

MACALLEN BUILDING CONDOMINIUMS

- Building type: Residential
- Location: Boston, Massachusetts
- Architect: Office dA, Ine. and Burt Hill
- Completion date: 2007

The Macallen Building is a condominium in southern Boston comprising 140 units. The condos were designed to incorporate and market a "green lifestyle." The building uses many different green technologies that are projected to save over 600,000 gal of water a year. It is expected that the building will use 30 percent less energy than a similar building not conforming to any specific green standards. The design team faced three main challenges: air and noise pollution, creating green space, and the urban heat island effect.

Green materials

- A minimal usage of steel was achieved by the staggered truss structural system.
- Rapidly renewable resources, such as bamboo, cork wallpaper, wood-fiber ceiling tile, and linoleum flooring were chosen.
- Seventy five percent of the wood is certified FSC standard.
- Various building materials, among which are concrete, steel, and aluminum siding, have recycled content.
- Materials were selected from sources no further than 500 miles away.

- Ninety percent of the construction and demolition waste was recycled. If a material could be reused, the team tried to find a new use for it; even drywall scraps were separated and reused.
- On each floor of the building there is a recycling system for residents.
- Materials chosen were formaldehyde-free and durable.
- Acoustical walls and ceilings are made of wood fiber.
- Excellent insulation was a major component of the building process, and the benefits in energy savings are being realized.
- Water-source heat pumps and heat-recovery ventilation technologies also save energy.

Green technologies

- Highly efficient water-source heat pumps control temperatures for the building.
- A cooling tower repels heat in the summer months; in the winter, a steam heat exchanger with steam generated at an off-site plant heats the building.
- Two heat wheels collect energy from exhaust, which later heats the building.
- Sensors control the lighting in areas not consistently in use, such as the garage and corridors.
- The sloped green roof has several important functions: it controls stormwater drainage; it reduces cooling and heating loads; it combats the heat island effect; it reduces heating and cooling costs; and it provides an ecosystem for vegetation.
- These same functions are also provided by a large outdoor terrace.

PHILADELPHIA FORENSIC SCIENCE CENTER

- Building type: Commercial
- Location: Philadelphia, Pennsylvania
- Architect: Cecil Baker and Croxton Collaborative
- Completion date: 2003

The Philadelphia Forensic Center made good use of the exterior structure of a former public school building, derelict for 20 years. Upgrades to the building envelope created an extremely well-insulated building. The diverse programs of the center include crime laboratories, a crime scene unit, chemistry laboratories for drug analysis, and a firearms unit. The facility is the first green building for the city municipal government. Because it is a laboratory, the Forensic Center has departments that demand very specific environments. Some spaces require great amounts of light and outside air, while others need to be completely contained in order to function properly.

Green materials

- Linoleum and agrifiber were used throughout the building for flooring purposes because they are ecologically friendly materials.
- Cellulose insulation was used to replace fiberglass batt insulation. Continuing the trend, cellulose carpeting was also used.

- The building is equipped with occupancy and temperature-based lighting and HVAC sensors to cut down on energy usage. What lighting is in use, unless otherwise specified by department, is high efficacy T8 fluorescents.
- A high-efficiency heating and cooling system was installed, achieving a reduction in energy use that paid for itself within three years.

Green technologies

- The massive window openings from the previous structure were used to advantage to achieve excellent daylighting.
- High R-value windows were of great importance, as this allowed for the fusion of new technology into the existing building envelope.
- Roof surfaces were designed with the ability to accommodate future photovoltaic systems.
- Ceiling configurations maximize daylighting.
- A very high standard of air quality has been achieved.
- The laboratory uses less than 50 percent of the energy a conventional forensic lab would require.
- Existing stairways and ventilation shafts were used.
- The existing underground spaces are reused for testing of firearms.

PITTSBURGH GLASS CENTER

- Building type: Commercial
- Location: Pittsburgh, Pennsylvania
- Architect: Davis Gardner Gannon Pope Architecture, LLC
- Completion date: 2002

The Pittsburgh Glass Center's purpose is to promote, create, and teach about glass. State-of-the-art studios now occupy a building that was formerly home to commercial establishments. The highlights of this building are the use of ecological materials inside and out, and the utilization of sustainable, high-performance technologies.

Green materials

- About 94 percent of the project's shell was reused, with some alterations to increase daylight and ventilation.
- Materials either came from the preexisting structure or were purchased from local businessmen who used recycled products.
- Special attention was paid to the recycled content of the new materials: salvaged wood was used; new wood was certified to be FSC qualified.
- Paint was almost totally avoided in the project. Generally speaking, the way the material arrived on-site was how it was used.

Green technologies

- The parking lot has pervious limestone rather than asphalt and is landscaped with native vegetation.

- HVAC systems include hot-water radiant-floor heating and 100 percent capacity fresh-air economizers.
- Digital controls monitor the heating and cooling systems, enabling the appropriate amount of energy to be expended.
- Occupancy sensors control the lighting in all spaces that are not consistently in use.
- The heat from the glass making process is recovered through coils and reused to control the indoor temperature of the building.
- Exposed concrete floors and ceilings control the thermal mass to help monitor the temperature.
- Only spaces such as the administrative offices, seminar rooms, and gallery have mechanical cooling. These spaces are insulated from the outside facade and are independent from the other systems of the building.
- The roof system is reflective, combating internal heat and urban heat-island effect.

SINO-ITALIAN ECOLOGICAL AND ENERGY EFFICIENT BUILDING (SIEEB)

- Building type: Commercial/Educational
- Location: Beijing, China
- Architect: Mario Cucinella Architects
- Completion date: 2006

Designed for the Tsinghua University in Beijing, the SIEEB is the marriage of two often-opposing goals: energy production and environmental protection. The building specifically aims at reducing CO_2 emissions while efficiently producing energy. The light and transparent architectural profile and the varieties of glass with different functions help achieve these goals. In a country where coal is still the major energy resource, this building and its programs look toward changing the future for the better.

Green materials
- Glass all around the façade has different types of finishes based on the direction it is facing.
- More than 1000 m^2 of photovoltaic panels on overhangs use the sun's energy to produce the primary energy for the building. It provides a natural way to power the facility while reducing the production of CO_2 gases, as well as providing shade for the terraces over which they hang.
- Gas engines along with electric generators supply supplemental energy if necessary.
- A green roof that helps filter the air through plants has been placed on top of the overhanging panels.

Green technologies
- The glass façades on this building serve a variety of functions.
- The south features a single skin glass that capitalizes on the transparency to let in light and heat. However, it is shaded by the projecting floors above.

- The north façade is designed to protect the building from cold winter winds. The glass is heavily insulated and mostly opaque.
- On the east and west sides the glass is double layered with horizontal sunshades.
- Sensor controls turn on lights only when there is movement in a particular area.
- This building uses a passive solar technology to provide green energy and to naturally light one façade while protecting its other façade from cold winter winds.

SOLAIRE RESIDENTIAL TOWER

- Building type: Residential
- Location: New York City, New York
- Architect: Building Green
- Completion date: 2003

The Solaire is a 293-unit, glass and brick tower, 27 stories high. Adjacent to the former World Trade Center (which delayed construction for a time), it is the first residential high-rise to obtain the gold LEED certification. The structure was designed to consume 35 percent less energy and to reduce water usage by 50 percent.

Green materials
- All materials are free of formaldehyde, with low VOCs, and 19 percent of them contain recycled material.
- The American Hydrotech Garden Roof system was chosen for both rooftops. This system starts first with a rubberized layer, then a root barrier, next a layer of insulation, fourthly a combination drainage/water retention/aeration layer, and lastly filter fabric. The plant species selected was bamboo, which adapts well to shallow soil and whose root systems can withstand high winds.
- Concrete containing fly ash, recycled-content gypsum board, brick (locally manufactured), stone, granite and slate, ceramic tile, recycled-content aluminum, and FSC certified wood were some of the green materials chosen.
- Photovoltaic panels contribute to energy production.

Green technologies
- Over 93 percent of construction waste was able to be recycled.
- Humidification and ventilation systems are installed at various levels, bringing fresh air to each unit.
- The thermal envelope, along with high-performance casement windows, maximizes daylighting.
- Programmable digital thermostats, Energy Star fixtures, and a master shut-off switch reduce energy use.
- Blackwater treatment on-site supplies the building's toilets and cooling system with water.
- A catch system provides water for the green roof.

STEINHUDE SEA RECREATION FACILITY

- Building type: Commercial
- Location: Steinhude, Germany
- Architect: Randall Stout Architects
- Completion date: 2000

The Steinhude Sea Recreation Facility is an advanced green building designed by Randall Stout Architects to serve as a visitor's center. Located on a natural island preserve in the middle of a glacial lake in northwest Germany, the facility is designed to blend into the environment of the island and leave as little impact as possible. Amenities for visitors include an exhibition area and observation deck, a cafe, and such necessities as lifeguard quarters and a boathouse. On an average the building serves 1000 people each week.

Because access to the island is difficult, sections were prefabricated in a factory and then erected in paneled modules by water cranes. The energy produced by the recreation facility is enough to power the building as well as recharge eight photovoltaic boats. The excess energy used is then sold back to the grid.

Green materials

- Recyclable concrete floor pavers, aluminum and glass curtain walls, and aluminum mullions are in line with the vision of minimal ecological impact.
- Mineral-fiber insulation is very environmentally sustainable since it is made from mineral wool and natural/synthetic metal compounds. It is very durable and provides good insulation (R-19 thermal value).
- Translucent polycarbonate panels (a type of thermoplastic polymer) reduce the weight of the building. They provide abundant light to the interior while protecting it from UV radiation. These exterior panels are also highly resistant to weathering and are more elastic and flexible than glass.
- Wood (including studs, plywood, glue-laminated beams, and timber framing) is FSC certified.
- Cement-paved units with sand-setting are used externally all around the center as a very effective type of erosion control. Excess water is allowed to flow through the units and the sand rather than creating runoff erosion.
- Photovoltaic panels provide shade for the roof and generate energy.

Green technologies

- A gray water collection system is used for bathroom facilities. Sewage water is drained by a pipe under the pedestrian boardwalk and from there to the sewer system on the mainland.
- Passive solar design in the form of location, orientation, and daylighting allows the building to use very little artificial light.
- Natural ventilation, instead of extensive HVAC systems, reduces carbon footprint and diminishes electricity consumption.
- A combined heat and power microturbine, fueled by rapeseed oil, is a backup system in case of overload or cloudy weather.

- Photovoltaic panels provide energy to the building, and solar hot water collectors minimize heating requirements.
- The building is automated for heating/cooling and lighting.

DISCOVERY CENTER AT SOUTH LAKE UNION

- Building type: Commercial/Community
- Location: Seattle, Washington
- Architect: The Miller/Hull Partnership
- Completion date: 2005

The Discovery Center at South Lake Union is a presentation center for the residential development and was used for exhibits featuring this neighborhood's "past, present, and future." As is generally the case, this type of structure is temporary and was built with this fact in mind. The Center sits on short concrete piers, and it was designed to be disassembled and moved to a new location. The cantilevered edges of the building allow the grade and vegetation beneath the facility to remain undisturbed.

Green materials

- Materials were selected with the idea that they must be easily demountable.
- Steel frames were shop-welded, painted, and then spliced together at the site.
- The roof framing system is based upon glued-laminated beam purlins and oriented strand-board sheathing. The roof was prefabricated in 8 by 20 ft sections and then moved to the site.
- Recycled concrete was used as a subbase in the parking area.
- Concrete containing fly ash composed the footings and piers.
- Exterior decks are made up of composite decking planks with recycled content.
- Paper-based composites made up the countertops, partitions, and exhibit display casework.
- Natural linoleum flooring and recycled carpet content was chosen.
- Large windows with an open layout allow the interior spaces to be filled with natural sunlight.

Green technologies

- Prefabrication allowed building time on-site to be greatly reduced.
- Four prefabricated modules make up the building, and they can break apart at the joints and be transported to a new site and possibly new program.
- High-efficiency toilets and waterless urinals save the building up to 15 percent of indoor water usage when compared to comparable buildings.
- Runoff reductions are due to rain gardens, planted swales, and porous asphalt or concrete.
- The four different modules have their own agendas for lighting, electrical, and mechanical systems. This allows the building to run more efficiently with different systems doing their own tasks instead of one large system that controls all of the building's uses.
- The building uses air-source heat pumps to cool and heat the building.

- The building has an EPA Energy Performance Rating of 60 and emits 12 percent less CO_2 than a baseline building.
- The site received a 97 percent diversion rate for all construction waste leaving the site.

GENZYME CENTER

- Building type: Commercial
- Location: Cambridge, Massachusetts
- Architect: Behnisch, Behnisch and Partners
- Completion date: 2003

Genzyme, a biotechnology company, wanted their center to be a symbol of progress—an innovative signature building that would identify them to their employees and visitors. The center's programs are offices for 920 people and space for the 480 visitors that arrive each week, besides training rooms, conference centers, a library, and a cafeteria.

The focal point of the structure is its grand 12-story central atrium. Open to the air, the atrium is not only visually impressive but also acts as a large return air duct and a light shaft. This alone yields a 42 percent savings in electricity costs.

Green materials

- Roof-mounted mirrors (heliostats) bring in natural light. They are completely automated, tracking the sun across the sky.
- Fixed mirrors then reflect the light to many prismatic louvers placed at the top of the atrium.
- Multiple hanging prismatic mobiles and a reflective light wall in turn continue the reflection of light all the way to ground level.
- The building envelope is a glazed curtain wall with operable windows. Thirty two percent of the façade is double ventilated and capable of both blocking and capturing solar gain.
- Air moving up the atrium leaves the building by exhaust fans near the skylight.
- Filigree concrete slabs are used for thermal efficiency as well as structural value. With the use of the concrete slabs, there was a total saving of 2552 cubic yards of concrete as well as 386 fewer tons of steel reinforcing bars. This also allowed the savings of 2600 sheets of plywood, significantly reducing the amount of VOCs released into the environment.
- Twenty three percent of the building materials are recycled and more than fifty percent were locally manufactured.
- Runoff reduction was established from a green roof system.

Green technologies

- Waterless urinals, sensors monitoring water use, and dual-flush toilets make a savings of 32 percent less water than comparable buildings. Rainwater is used for the evaporative cooling towers as well as for irrigation for the green roof system.
- A local power plant powers the central heating and cooling system of the building with the use of steam. The steam drives absorption chillers for cooling and can be

directly exchanged into heat for the winter season. This allows the building to avoid such distribution losses and keeps the electrical usage low during peak summer electrical demands because of the use of heat to operate instead of electricity.

- Fan coils are used to pump water throughout the building instead of blowing air, meeting heating and cooling loads in each space.

Figure 8.14   One external view and two internal views of the Genzyme Center.
(Images courtesy of Behnisch, Behnisch and Partners)

- A series of sensors placed around the building allow for control of lighting, ventilation, and temperature.
- Employees are encouraged to use bicycle or public transportation.
- Temperature controls can be individually set, allowing workers to control their own space.
- Two photovoltaic arrays generate 20 kW at peak output.

HAWAII GATEWAY ENERGY CENTER

- Building type: Assembly/Interpretive
- Location: Kailua-Kona, Hawaii
- Architect: Ferraro Choi and Associates, Ltd.
- Completion date: 2005

The Hawaii Gateway Energy Center (HGEC) is an interpretive visitors' center that offers displays, education, outreach programs, and laboratory research. It educates the public about future potential for solar power and efficient buildings. Its 20-kW solar array, which generates 24,455 kWh of electricity a year, is the most outstanding feature of the building. The structure is designed to be a huge thermal chimney, capturing heat and creating air movement.

Green materials

- Concrete and lava rock were obtained from production plants no more than 25 miles away.
- Steel, copper roofing, insulation, carpet, and flooring are some of the materials containing recycled content.
- Carpet tile replaced glue-down carpet in the multipurpose space.
- Thermoplastic olefin (TPO) roofing and copper roofing radiate heat from the sun into a ceiling plenum. As the heated air rises it is exhausted through stacks.
- The ceilings are of acoustic ceiling planks.
- The walls are of thermal and acoustical insulation and gypsum board.
- Gravity-laid resilient flooring was chosen for the administrative areas.
- A polyethylene sheet of 100 percent postconsumer recycled resin is a low VOC special surfacing.

Green technologies

- Proper orientation and design, along with glazing, makes the use of electric lighting almost unnecessary.
- Cold seawater is pumped in and distributed through cooling coils, passively cooling the building. The only energy needed is enough to run the seawater pump.
- Condensation that drips off the cooling coils is used for flushing toilets and irrigating.
- The building was designed with an on-site, 20-kW photovoltaic array, providing nearly half the energy required for the building's operation.

- Only about 20 percent of the energy that would be required by a comparable building is needed.
- The excellent energy efficiency makes the HGEC a net-zero-energy building that actually exports electricity.
- The occupants practice an in-building recycling program.

LAVIN BERNICK CENTER FOR UNIVERSITY LIFE

- Building type: Commercial/Educational
- Location: New Orleans, Louisiana
- Architect: Vincent James & Associates
- Completion date: 2007

The Lavin Bernick Center for University Life is located on the Tulane University campus and is a nucleus for students and extracurricular activities. Its program includes a bookstore, dining facilities, study areas, and meeting rooms. The structure is built on the same site as the preexisting center, which was demolished and redesigned to be more environmentally friendly. The center is typically occupied by 110 people with perhaps 7000 visitors each week. Balconies, canopies, shading systems, and courtyards pay homage to the culture of New Orleans and reduce solar gain as well.

Green materials

- Cladding materials on the exterior were chosen for durability and low maintenance.
- Cladding materials on the interior were chosen for their longevity and thermal performance.
- Terrazzo floors use thermal mass to regulate temperatures.
- The microporous ceiling is metal that diffuses dehumidified preconditioned air and is also an acoustical surface.
- Large oak trees were preserved to facilitate shading courtyards and building.

Green technologies

- The existing concrete structure was reused.
- The building is cooled by both natural ventilation and mechanical cooling; it is allowed to stand open to the exterior when weather permits, decreasing cooling requirements by 42 percent.
- Extensive shading systems on the exterior allow the building to take advantage of as much natural daylight and fresh air as possible.
- Energy consumption is reduced with standard cooling only in regulated "thermal refuge" zones.
- The building envelope used insulation, increased glazing, and shading to further reduce annual energy consumption.
- The center's orientation allows natural breezes to cool spaces.
- Low flow toilets contribute to water conservation.

- As the building sits below the water table, flooding is a common problem. The site is designed to allow percolation into the soil or transfer water into the city canal system. Permeable paving and drainage systems are installed.
- Many light wells at the garden level let natural light into the basement, helping to reduce energy consumption.

LIGHTHOUSE TOWER

- Building type: Commercial
- Location: Dubai, United Arab Emirates
- Architect: Atkins, Hill International Architects
- Completion date: Estimated 2010

The Lighthouse Tower will be built for the Dubai International Financial center as a mixed use development project. The primary goal of the project is a sensitive approach to the environment as well, and energy efficiency and reduced carbon emission techniques will figure highly in the design. In pursuit of this goal the building will be outfitted with 4000 photovoltaic panels, and three huge wind turbines. It will be 66 stories tall, used primarily for luxury office suites, but will also include parking, a convention center, retail outlets, a restaurant, gym and swimming pool, and gardens on the ground level and throughout the upper floors of the skyscraper.

Green materials

- All materials are to be selected from sustainable sources.
- Green tinted glass keeps energy consumption down by reflecting intense sunlight away from the building, thus contributing to cooler interior temperatures. In addition, the use of glass curtain walls optimizes daylighting, reducing electricity consumption.
- White aluminum panels work in the same way as the tinted glass, reflecting sunlight and reducing the need for air conditioning inside. In addition, aluminum is a highly recyclable material.
- Reinforced concrete, within the context of this building, is a sustainable material; because of its durability it promises a long life span for the building.
- Steel will be used to frame the upper portion of the Lighthouse Tower. This will increase the height of the building without a dramatic consumption of resources.
- Environmentally friendly application agents such as paints, coatings, adhesives, sealants, etc., will be used.

Green technologies

- The company has pledged activity pollution prevention during construction.
- Four thousand photovoltaic panels on the south facade will provide energy to the building to offset its dependence on grid energy.
- Three wind turbine power generators will crown the tower and rotate to generate electricity.

- Actively cooled beams will combat the intense heat of Dubai and reduce the need for internal air-conditioning.
- In combination with passive daylighting techniques, motion sensors will activate and deactivate lights, conserving energy when a space is unoccupied.
- The building is planned to offset its own energy consumption by 50 percent and its water consumption by 40 percent.
- An in-house system of storage and collection of recyclables is promised.
- Tap and shower flow restrictors, waterless urinals, and low-flushing toilets will reduce water consumption greatly by restricting the amount of water used for daily hygienic activities.
- Gray water recycling into irrigation will make good use of wastewater while preventing potable water from being used for irrigation.
- A seven-level parking garage is an efficient use of space. Stacking parking up instead of out uses less area to fulfill a program by conserving spatial resources.

SWISS RE TOWER

- Building type: Commercial
- Location: London, England
- Architect: Foster and Partners
- Completion date: 2004

The Swiss Re Tower of London was designed for the Swiss Re, a global reinsurance company, for their operations in the United Kingdom. It is said to be London's first environmentally conscious skyscraper. The building primarily contains offices of the Swiss Re but includes smaller, local businesses. It also contains several levels for eating, including a bar that sits on the top floor (40th level) which offers a 360° view of London. Its distinctive shape, which does indeed resemble a gherkin or cucumber standing on end, has spiraling bands around it. Its aerodynamic shape minimizes wind loads while providing outstanding light, ventilation, and views.

Green materials
- Spiraling bands around the building allow for the building to obtain large amounts of lighting to penetrate into the workspaces to make the environment more enjoyable. These bands also are the dynamic of the ventilation system which can relay 40 percent of the energy from the mechanical ventilation system to be shut off throughout the year.
- Six shafts serve as a ventilation system. The shafts form a double glazing effect; between the two layers of glazing is a layer of air which insulates the interior.
- Gardens at every sixth floor purify moving air.
- The structure is made of steel in triangular forms, which is so strong that no columns are needed for support on the interior.
- Over the steel structure are flat diamond-shaped glass panels, varying in size at each level.

Green technologies

- The façade of the building is able to open to allow fresh air flow throughout the building, reducing air-conditioning usage.
- Glazing on the exterior of the windows allows refraction of the sunlight in the summer and harnessing of the sunlight in the winter when heat is needed for the building.
- Gaps at every six levels allow for ventilation throughout, allowing less energy usage as well as fresh air in the work areas for better, more productive working conditions.
- Gas is the primary fuel for the building.
- Sensors are placed throughout the building to control lighting. Movement sensors that monitor each room keep lights on when in use and off when no one is occupying the rooms.
- Only one level of parking is available for the workers, which lowers the incentive of driving personal vehicles to the site and encourages occupants to take other means of transportation such as the Liverpool street station. The building also offers 3 times as much bicycle parking as comparable structures.
- The unique structure of the building allows more load-bearing strength but less material to be used within.
- The distinctive shape responds to the size of the site, deflects wind, and does not overshadow pedestrians at its base.
- The building consumes only 50 percent of the energy of a traditional office building.

WAYNE L. MORSE UNITED STATES COURTHOUSE

- Building type: Government/Public
- Location: Eugene, Oregon
- Architect: Morphosis
- Completion date: 2006

The Wayne Lyman Morse United States Courthouse is part of the Ninth Judicial Circuit in Oregon. The courthouse is a five-story building, of which one floor is a basement, with six courtrooms and offices. Because the courthouse is at high risk for attacks and the need for security is paramount, the security system was a dynamic factor in the total design. Complex and strong security systems, but with the use of local and ecofriendly materials, were installed. And although Judge Hogan preferred a traditional style, what he got was a modern steel and glass structure that, he admits, meets all his specifications.

Green materials

- More than 20 percent of the building's materials are made up of recycled content.
- Steel and aluminum components, including rebar and structural steel all have recycled material.

- PVC-free interior shade screening was used throughout the building to reduce cooling needs.
- Materials with low levels of volatile organic compounds (VOCs) were chosen to produce and maintain good indoor quality within the courthouse walls.

Green technologies
- The building's energy use was reduced by approximately 40 percent through the use of daylighting, high performance glazing, ventilation, and radiant-floor heating and cooling.
- Daylight and occupancy sensors help to control the fluorescent lighting throughout spaces.
- Evening air replaces daytime air in the building each day to help reduce cooling loads.
- Waterless urinals, low-flow toilets, faucets, and showerheads reduce water use by more than 40 percent when compared with a conventional facility.
- The underground parking infrastructure helps to reduce the building's overall footprint and minimize massive expanses of asphalt.
- Native drought-tolerant plants can survive with natural precipitation, reducing water used for landscaping by approximately 60 percent.
- The irrigation system takes advantage of rainwater.
- The radiant flooring system throughout the building helps to heat the large spaces without massive amounts of energy.
- Shading structures shelter southern windows from solar heat.

WILLIAM J. CLINTON PRESIDENTIAL LIBRARY AND MUSEUM
- Building type: Educational/Public
- Location: Little Rock, Arkansas
- Architect: Polk Stanley Yeary, Witsell Evans & Rasco and Woods Caradine
- Completion date: 2004

The William J. Clinton Presidential Center sits within a new, landscaped public park along the south bank of the Arkansas River. The building has a bridge-like form, cantilevered over the river and reminiscent of Little Rock's iconic six bridges. Symbolically it echoes Clinton's pledge to "build a bridge to the 21st century." Information about Clinton's presidency is found in a large, daylit exhibition space, while archives are found in a below-grade secure environment.

Green materials
- Materials all had to pass the test of local availability, recycled content, rapidly renewable content, and/or low chemical emission properties.
- A screened interlayer on the west wall of glass blocks half of the sun's heat and 99 percent of the UV rays.

- Material from preexisting buildings was reused; for instance, the old Choctaw Station is now part of the University of Arkansas School of Public Service and an abandoned railroad bridge became a pedestrian bridge.
- Demand-controlled ventilation and radiant-floor heating and cooling is installed.
- Some electricity is produced by the 50-kW photovoltaic array.
- Two levels feature bamboo flooring, a highly renewable resource.
- Ninety five percent of the Center's waste is recycled.
- Recently green cleaning recycling programs were added, along with water reducing landscaping and a green roof.

Green technologies
- The Center uses 34 percent less energy than comparable code compliant buildings.
- Three hundred and six solar panels provide part of the building's energy.
- The Center uses 23 percent less potable water than a comparable building.
- Ninety nine percent green cleaning products are used.
- Carefully selected screens and glass reduce solar heat gain by 50 percent.
- Transportation other than automobile is encouraged via bicycle racks, shower, and changing rooms. Priority parking is given to buses and charging stations are available for electric vehicles.

YORK UNIVERSITY COMPUTER SCIENCE BUILDING

- Building type: Educational/Public
- Location: Toronto, Canada
- Architect: Busby and Associates/Architects Alliance (formerly Van Nostrand Di Castri Architects)
- Completion date: 2001

The York University Computer Science Building was the first green building in Ontario when it was completed in 2001. The challenge was to design an exposed building that received energy from solar gain and heat absorption in a cold-weather climate. Three zones—a three-story bar building, a courtyard building, and a lecture hall—comprise the structure. An atrium planted with bamboo links two zones and aids in circulation. Extensive research went into making the building acoustically sealed, using acoustic lining. Above is a canopy of refined steel and glass that provides weather protection.

Green materials
- Copper façade paneling, a highly recyclable material, is used on the facades in place of a nonrenewable material.
- Fifty percent fly ash concrete has a very long life span.
- Postrecycled steel is sustainable and durable.
- Reused timbers were used in part of the building.
- Permeable walkways have a much smaller impact on the site's hydrology and vegetation than standard concrete sidewalks.

Green technologies

■ The canopy over the courtyard allows for passive solar heating, based on the same principles as those of a green house.

■ Thermal chimneys regulate interior temperatures by venting overheated air during the warmer months.

■ Operable sunshades regulate sun lighting and active solar heating.

■ A planted roof with water collection tank cools the building and reduces water runoff.

■ Saw-toothed façades use copper to redirect solar energy, thus reducing heating and cooling costs.

■ Energy consumption is 50 percent less than a comparable building.

Active Solar Buildings

BIO-SOLAR HOUSE

■ Building type: Residential
■ Location: Bangkok, Thailand
■ Architect: Soontorn Boonyatikam
■ Completion date: 2006

The Bangkok Bio-Solar house is a self-sustaining adobe, designed by Boonyatikam, a university professor of architecture at Chulalongkorn University. Described as "an environmental dream," the house generates its own electricity, water, and cooking gas. One hundred percent of its own power is produced through solar panels, while its water supply is maintained through rain, dew, and condensation from the cooling system. Enough excess energy is produced to power an electric car 30 miles a day (or sell back to the electric company). At first glance hardly memorable, the house's most outstanding feature is a "green room," a square, glass-encased space (even the floor is of glass) cantilevered out over a swimming pool. The house is completely sustainable and ecological. It cost about $75,000 to build.

The house, which is virtually airtight, also helps with pollution by filtering the air in the house, then pumping fresh air out. This house, to an extent, is part of nature because what it takes out, it gives back, cleaner and more sanitary.

Green materials

■ Overhanging eaves and recessed windows limit heat gain and glare. Windows and doors have triple-paned glass.

■ A metal roof covers an air duct, which allows wind to ventilate the roof, and insulation 1 ft thick.

Green technologies

■ Six hundred and seventy square feet of solar cells mounted on the roof are used to produce energy for day-to-day use, such as air-conditioning, lighting, and household appliances. The system is capable of storing energy for three days, and it is 14 times more energy efficient than a traditional house.

- A personal computer linked to sensors around the house controls the system, like regulating temperature and humidity.
- Condensation from the cooling system along with natural moisture from rain and dew is used to collect all the water needed for the house and landscaping.
- Recycled water is used for irrigating the garden.
- Special filters are used to clean the air coming in and out of the building, which is especially important in Bangkok's polluted atmosphere.
- The house is self-sustaining, which means no utility bills. It should pay for itself in seven years.

EASTERN SIERRA HOUSE

- Building type: Residential
- Location: Gardnerville, Nevada
- Architect: Arkin Tilt Architects
- Completion date: 2004

Nearly energy independent, this green demonstration house built in the Sierra Nevada Mountains, blends with the landscape as well as complements it. Centered around an "oasis courtyard" shaded by photovoltaic panels, the design showcases the use of new technologies and green practices. The site was carefully considered to take advantage of the slope and dramatic views of mountains.

Green materials

- The photovoltaic panel array provides power to the house as well as acting as a shading element. It is an effective, relatively economical, and low maintenance system.
- Ten inch structural insulated panels (SIPs) make up the roofs and have an R-45 insulation value.
- The walls are constructed of FSC wood, sheathed with fiber cement board panels.
- Four foot overhangs are of salvaged fir.
- Salvaged materials, such as beams and shade fins from an old airport, are used throughout the house.
- Native species and a permaculture garden promote local earth resources.
- Straw-bale construction with PISE finish.

Green technologies

- Solar collectors heat water for interior use and space heating.
- SIPs are used for the roofs, including the living room.
- Hot water travels into loops of tubing underground beneath the living area. The thermal energy radiates throughout the interior.
- Rainwater is collected in more than one location and then diverted to swales.
- Local wildlife uses the swales for watering holes. Irrigation from the swales helps maintain native habitat zones with several edible plants.

THE GEMINI HAUS

- Building type: Prototype
- Location: Weiz, Germany
- Architect/Inventor: Roland Mosl
- Completion date: 2001

The Gemini Haus uses the technology that Architect/Inventor Mosl invented in 1993, and after an additional eight years made a reality in his prototype building in Weiz, Germany. The key to the technology is the photovoltaic solar panels that wrap around the entire house, taking advantage of the sun at every moment of the day. The round design allows the solar equipment to rotate with the sun; thus there is no need to struggle for solar gain. The solar panels are the main focal point of this prototype.

Green materials

- The house is partially wrapped in glass to maximize natural daylighting and reduce the need for artificial lighting.
- Extreme thermal insulation maintains temperature.

Green technologies

- The heat recovery system helps prevent energy loss.
- One hundred and fifty square meter of turning photovoltaic panels, which provide thermal insulation as well as energy to run the house, rotate to take advantage of the sun.
- The house creates enough energy to produce a surplus.

GEOTHERMAL HOUSE

- Building type: Residential
- Location: Boston, Massachusetts
- Architect: MaryAnn Thompson Architects
- Completion date: 1997

Geothermal House is a residence that relates the interior to the exterior. Design factors include horizontal planes that form terraces along a hill overlooking a pond. The house maintains a low profile and windows provide a "hide and reveal" of the landscape. All rooms get daylight from two sides, and several rooms receive light from all four sides by way of a clerestory.

Green materials

- Large windows open up the house to the outdoors.
- The architect wished to site the house so that the sun would be an "ever-changing presence" in the house.
- The kitchen faces east, which takes advantage of the sun in the morning; the living room and terrace face west, taking in the evening sun.

Figure 8.15 Two views of the Gemini Haus. *(Images courtesy of Roland Mosl)*

- Since the house lets in so much light, the family barely uses the heater during the winter.
- The western facade takes advantage of overhanging trellises.

Green technologies
- Windows are designed to permit all rooms to have cross ventilation.
- The arrangement of the house is based on topography, solar orientation, and views.
- Heating and cooling system are geothermal, which saves 60 percent in energy costs.
- Overhangs provide shade in the summer but light in the winter.

REVITALIZED SOLAR-POWERED UNION LOFTS

- Building type: Low-Income Housing
- Location: San Diego, California
- Architect: Jonathan Segal Architecture
- Completion date: 2006

The revitalized solar-powered union lofts are a complete reuse of an old industrial warehouse and other commercial buildings. It is a stunning 16-unit housing complex that is affordable and sustainable. The complex takes up a 20,000-ft^2 block. When the construction was completed, it achieved a number of awards from AIA for being an accomplished green building and was described as "a paragon of efficient modern minimalism."

Green materials
- Roof-mounted photovoltaic panels produce 50 percent of the necessary energy.
- Each unit has the advantage of plenty of natural light and ventilation.
- A private patio forms a part of each unit.

Green technologies
- The photovoltaic panels have no adverse effect on the environment.
- Natural ventilation uses no energy and produces a cool interior.
- Adaptive reuse in this case meant using the original buildings of a warehouse, convenience store, a gas station, and an old textile building.
- Steel structures are not necessarily beneficial to the environment, yet its strength gives tremendous sustainability.
- Concrete provides the building with great stability while having minimal impact.

SANYO SOLAR ARK

- Building type: Active Solar, Museum/Assembly
- Location: Gifu, Japan
- Architect: Sanyo Design Team
- Completion date: 2001

Figure 8.16 Views of the Union Lofts from four different sides.

(Images courtesy of Paul Body)

Figure 8.16   *(Continued)*

The Sanyo Solar Ark is aptly named, as it strongly resembles a modern Noah's ark. It collects an impressive 630 kW with over 5000 solar panels. An amazing 500,000 kWh of energy per year is produced. It is an outstanding example of building with integrated photovoltaic design. Multicolored lighting between the panels can create a variety of letters or shapes on the sides of this enormous structure. The unique building houses a solar museum and solar lab.

Green materials

- Nearly all of the monocrystalline modules on the photovoltaic were saved from landfills, where they were headed due to their insufficient output, an example of "turning lemons into lemonade."
- LED lighting is used on the interior and exterior and is computer controlled; thus, it is only activated when needed. LEDs have a long life, which reduces waste when compared to traditional lighting configurations and use very little energy.
- Solight is a unique lighting apparatus equipped with a compact motor driven by a small solar cell. It automatically changes its direction as the sun moves in order to receive the optimum sunlight, which is then used as a natural light source. This sustains the life of the lighting system. Solights produce light naturally, thus reducing CO_2 emissions. They perform as an entire system for interior lighting and are extremely efficient.

Green technologies

- Two ponds located on the site of the building double as a fire hydrant reserve. The ponds are cleaned with the SANYO Aqua Clean system that electrolyzes the water to produce hypochlorous acid, cleaning the water continuously. The cleaning system is all-natural, requiring no chemicals.
- Recycled solar panels do double duty as a technology because they produce a large amount of energy as well as perform as the exterior cladding system.
- A total CO_2 reduction of 95 tons per year has been achieved.
- The building's curve takes advantage of the sun's light and heat.
- A truss system throws out cantilevers from the building's center.

SOLAR TUBE

- Building type: Residential
- Location: Vienna, Austria
- Architect: Driendl Architects
- Completion date: 2001

A solar tube is a small light and heat-capturing device that is typically installed on the roofs of high-efficiency homes. However, the Solar Tube House is virtually a huge solar tube itself. Central to the home is a three-story atrium which acts as a ventilation shaft and naturally controls the climate. Mature trees arching over the structure add their own protection with a combination of shade and filtered sunlight. The house has a central core of reinforced concrete, which both absorbs and stores the warmth; this process (called thermal mass) helps keep the house at the desired temperature. The house really gets "passively aggressive about energy."

Green materials

- The windows contain a layer of metal sandwiched between two sheets of glass that absorbs the short warming rays but deflects the longer, damaging UV rays.
- Locally produced maple and walnut wood make up flooring and doors.
- Granite from a local quarry helps reduce the emissions associated with transportation.
- Mostly prefabricated units were used, allowing the house to be completed in only five months.

Green technologies

- A strong commitment to solar energy and heating is the basic principal Driendl followed while creating this house.
- Attention was paid to solar orientation while designing the building.
- Insulation materials that are known to collect and store heat were used.
- Operable windows aid cooling in the summer and retain heat in the winter.
- The curtain wall traps and stores energy as the material absorbs radiant heat.

Figure 8.17   Solar Tube. *(Images courtesy of James Morris (top) and Bruno Klomfar (center and bottom))*

Figure 8.17   (*Continued*)

SOLAR UMBRELLA HOUSE

- Building type: Residential
- Location: Venice, California
- Architect: Pugh and Scarpa
- Completion date: 2005

The Solar Umbrella House, originally built in 1923, has undergone two renovations, one in 1934 and the latest in 2005. In this latest renovation, 1150 ft² were added. Lawrence Scarpa, architect and owner of the dwelling, ignored advice to demolish the building and decided to renovate, add on, reuse, and reinforce.

The most distinctive design feature of the Solar Umbrella House is the shading solar canopy. However, the canopy does not just have the usual task of providing shade; it uses its 89 photovoltaic panels to transform sunlight into energy, providing 95 percent of the total.

Green materials

- The concrete floors, which are 50 percent fly ash, supply heat through an integrated solar heating system.
- Lighting control systems were installed for both interior and exterior lights.

- Recycled mild steel, rusted and then sealed, is very durable.
- All the paint is low-VOC.
- Cellulose insulation achieves an R-value of more than 15.
- High-performance windows and doors are used throughout the house.
- Industrial broom bristles are used for vertical shading.
- Exterior materials are natural pigmented stucco, concrete, and recycled and rusted steel, none of which requires painting.

Green technologies

- All of the concrete forms and 20 percent of the framing materials were reclaimed.
- Rooftop solar hot-water panels heat the pool and preheat the domestic hot water before it gets to the gas-fired water heater. Overall, this system reduced the natural gas use by half.
- An open floor plan allows exterior light deep into the interior.
- The construction team recycled over 85 percent of construction waste.
- A drip-irrigation system provides water to the native landscaping.
- A stormwater retention system retains 80 percent on-site.
- This project maintains over 65 percent of the 4100-ft^2 site as landscaped or not paved, reducing both runoff and the project's contribution to the heat-island effect. It leaves a very minimal footprint.

WINE CREEK ROAD HOME

- Building type: Residential
- Location: Healdsburg, California
- Architect: Siegel and Strain Architects
- Completion date: 2002

The Wine Creek Road home is in keeping with local architecture and was built as a summer retreat. Since the area has hot summer weather, one of the main goals was to keep the house cool, but to use only natural ventilation to limit environmental impact. The floor plan divides the building into two main areas separated by an open dogtrot/breezeway. On one end is the living and eating area; on the opposite end are the bath and bedrooms.

Green materials

- Straw bales provide low-tech, high-performance insulation which keeps the house cool. They are natural and nonpolluting.
- The placement of the double-glazed windows supports a very efficient ventilation method.
- The plaster walls and concrete floors, which cool down at night when windows are opened, keep the house cool the next day.
- Radiant floor heating is provided by a water heater.

Green technologies

■ The structure of the house is a combination of straw bales, cellulose insulation, and framing.

■ Straw bales used as walls have excellent thermal qualities. Their mass allows them to absorb heat during the day and reradiate it at night.

Figure 8.18   Two views of the Wine Creek Road Home along with a graphic representation. *(Images courtesy of J.D. Peterson (top and center) and Siegel & Strain (bottom))*

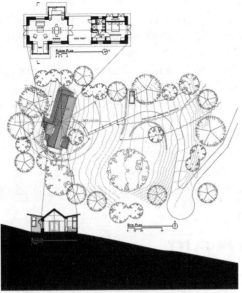

Figure 8.18   (*Continued*)

■ The breezeway is a fundamental design element which is regaining popularity. By careful positioning of the house, it captures prevailing air movement and funnels it through a space, effectively cooling even when the air itself is warm.

Z6 HOUSE

■ Building type: Residential
■ Location: Santa Monica, California
■ Architect: Ray Kappe, FAIA Kappe Architects/Planners
■ Completion date: 2006

The Z6 home has earned its title, as it has attained zero levels of six environmental hazards: "waste, energy, water, carbon, emissions, and ignorance." The overarching goal was to use every possible green building method to make a zero-impact dwelling. The house is not only a residence but a model home for similar buildings that the owner constructs.

Green materials
■ Low-emissivity glass with special glazing reduces heat transfer through thermal conductivity.
■ A photovoltaic system along with a solar hot water heater provides clean and natural energy to power the home.
■ The green roof reduces runoff and a water cistern collects rainwater for irrigation.
■ Cork floors are practical and can be harvested without harming the tree.
■ Recycled products were used for everything from tiles to kitchen countertops.

Green technologies

■ The house was built from prefabricated modules which were transported to the site. Prefabrication allows for the more efficient use of materials. Setup took no more than 13 h.

■ Cooling the house in this area needs nothing more than cross ventilation and a thermal chimney.

■ Heating is achieved simply through radiant floor heat powered by a solar hot-water collector.

■ The photovoltaic array is backed up by a battery storage system which ensures that even during a blackout, the home will be powered.

■ Gray water irrigates the native (and edible) plants.

■ Flexible use of space is derived from the movable wall partitions, allowing the owners to create spaces that best suit their needs.

Z-SQUARED (Z2) DESIGN FACILITY

■ Building type: Commercial
■ Location: San Jose, California
■ Architect: Integrated Design Associates Inc. (IDeAs)
■ Completion date: 2007

The Z-Squared building produces zero carbon dioxide emissions and uses zero energy. "This was a big issue for us, to show that it is possible now to do a zero carbon building," David Kaneda, president of IDeAs says. "It is not a dream of the future, but a dream that is attainable right now." Before renovation, the Z-Squared building was windowless—a block of solid concrete. In the 1960s, it was a bank building in San Jose. The thrust was to design it around the goal of energy efficiency, and the firm has succeeded.

Green materials

■ A photosensor automatically tints the windows, made with electrochromic glass, to reduce glare and heat transfer.

■ UV filtering glass makes up the skylights and sliding glass doors.

■ A heat pump forces warm or cool water through underfloor polyethylene tubes to heat and cool the interior.

Green technologies

■ Solar power comes from skylights and photovoltaic panels.

■ Equipment when in sleep mode still uses current. The security system now can send a signal to the electrical equipment which switches off the circuit breaker.

■ The use of high-efficiency office equipment and automatic controls minimizes the load put on plugs.

■ Sensors monitor energy use in the building.

■ The building keeps a stable temperature with bioinsulation.

DATAGROUP OFFICE BUILDING

- Building type: Commercial
- Location: Pliezhausen, Germany
- Architect: KTP (Theilig Kauffmann and Partners)
- Completion date: 1995

The Datagroup Office Building, whose program is primarily office space for software development, is a round structure, enabling most workers to enjoy the outstanding views of the surrounding Swabian Alb mountains. The building slopes down in five levels, corresponding to the terrain. The design focuses on being environmentally friendly, as well as on being conducive to productivity and collaboration among employees.

Green features of the building focus on an active approach to solar heating. The building is arranged radially, placing offices around a central atrium space. This allows heating and cooling to be more efficient due to the cohesiveness of the design. The atrium roof is of glass, with a fabric "flower" cover that responds to light, opening and closing depending on the time of day. This active approach to solar heating and cooling works to decrease the energy consumption of the building.

Green materials

- Within the context of this building, glass is a sustainable material since it is the main component of the building's active solar design, which in turn reduces energy consumption.
- Cable is used to create an atrium roof that is much simpler than the extensive roof assembly that would otherwise be used, sustainable in that it consumes less material than equivalent alternatives.
- The fabric used to control daylighting in the building's atrium is considered ecological since it is a low-impact material that can be reused.
- Relaxation breaks can be enjoyed by the workers on large wooden balconies overlooking the countryside.

Green technologies

- Light responsive atrium screen: Cables, moving parts, and light-detecting technology work together to create the active solar system in the atrium. The giant fabric "flower" opens and closes depending on the amount of light coming into the building to provide optimum solar heat gain.
- Computer stations are shielded from direct sunlight by room dividers, some with skylight strips, so only northern or indirect light falls on the monitors.
- Heat derived from people, monitors, scanners, printers, etc. produce enough waste heat that even in winter a heater is unnecessary.
- In summer, this waste heat is so intense that natural ventilation is not sufficient. Cooled air is first forced into the atrium and from there through the vented corridor walls and into the offices. The return air is collected by an air duct running along the exterior of the building.
- Before air enters the building it is transported to an underground duct that is deep enough in the earth to reduce the temperature of the air a great deal by natural means.

Figure 8.19   Two external views and one internal view of the Datagroup Office
Building. *(Images courtesy of Roland Halbe)*

Passive Solar Buildings

MASDAR HEADQUARTERS

- Building type: Commercial
- Location: Masdar City, Abu Dhabi
- Architect: Adrian Smith + Gordon Gill Architects
- Completion date: Estimated 2010

Masdar Headquarters is a 1.6 million ft^2, zero-waste, zero-carbon emission commercial complex. A huge canopy lined with photovoltaics is supported by eleven massive glass hyperboloids. The canopy shelters offices, stores, and residences. A reduction in energy is planned from the start when the photovoltaic canopy is installed first. The structure will then shade the workers as it turns sunlight into energy for the project.

Every available green technology and material has been incorporated into the project, including green roofs, air quality sensors, and wind turbines. The structure will use 70 percent less water than a similar building of its size.

Green materials

- Local and recyclable materials.
- Ecological materials with minimal waste.
- Outdoor air quality monitors to monitor local air pollution. Green roofs are used to provide natural insulation, to produce food, and to recycle food waste.

Green technologies

- Glass hyperboloids, which support the canopy, are like "cooling chimneys"; they are used to exhaust warm air and also create interior courts with water features. They prevent glare and bring in diffused light. The world's largest cooling and dehumidifying system is driven by solar power.
- Building integrated photovoltaic arrays produce more energy than the building consumes.
- Wind turbines are used to produce pollution-free energy.

CARNEGIE INSTITUTION OF WASHINGTON GLOBAL ECOLOGY CENTER

- Building type: Commercial/Education
- Location: Stanford, California
- Architect: Estherick, Homsey, Dodge, and Davis—EHDD Architecture
- Completion date: 2004

The Global Ecology Center is located on the campus of Stanford University. It is an extremely low-energy laboratory and also serves as a university office building. Projects and research in the facility center around the interactions between the earth's ecosystems, land, atmosphere, and oceans. The school wanted the architects to focus on reducing carbon impacts and water use and address biodiversity, while at the same time providing laboratory and research spaces that meet the highest standards of green architecture. The designers also pursued habitat and water conservation goals.

Green materials

- Salvaged wine-cask redwood was used for the exterior.
- Tables were constructed from salvaged trees or salvaged doors from a nearby location.
- No-irrigation landscaping, dual-flush toilets, waterless urinals, and low-flow sinks cut water use by one-third.
- FSC-certified ash makes up the interior wood.
- High-volume fly ash concrete: Foundation, sidewalks.
- Twenty percent of the site's concrete aggregate is recycled.
- Use of green building materials resulted in a 50 percent reduction in embodied carbon.

Green technologies

- A hydronic system cools down the building by spraying water on the building at night. The night air cools the water, which runs down the roof, losing heat as it goes. The water is stored in an insulated tank which is used in cooling the building.
- Stormwater is recovered and reused on-site.
- A cooling tower uses downdrafts to refresh the interior air of the lobby and is an iconic landmark.
- On windless days, atomizing spray nozzles in the tower cool the air, creating a thermally driven downdraft in the lobby.
- Green practices achieved a 72 percent reduction of CO_2 emissions.
- Daylighting, sunshading, and natural ventilation lower the use of electricity.
- Occupancy sensors and photosensors dim lights when there is enough daylighting and only light a room when occupied.

NUEVA HILLSIDE LEARNING COMPLEX

- Building type: Education/Public
- Location: Hillsborough, California
- Architects: Leddy Maytum Stacy Architects
- Completion date: 2007

Three hundred and seventy students attend the private school Nueva Hillside Learning Complex, which consists of three buildings surrounding a center courtyard: a library, a student center, and a classroom building for grades 5 to 8 with administrative offices. It is built on the former site of a large parking lot on the school's campus, which negated the necessity of tearing down existing buildings. All the materials were selected for economy, durability, and resource efficiency. Eighty percent of all the construction debris was recycled.

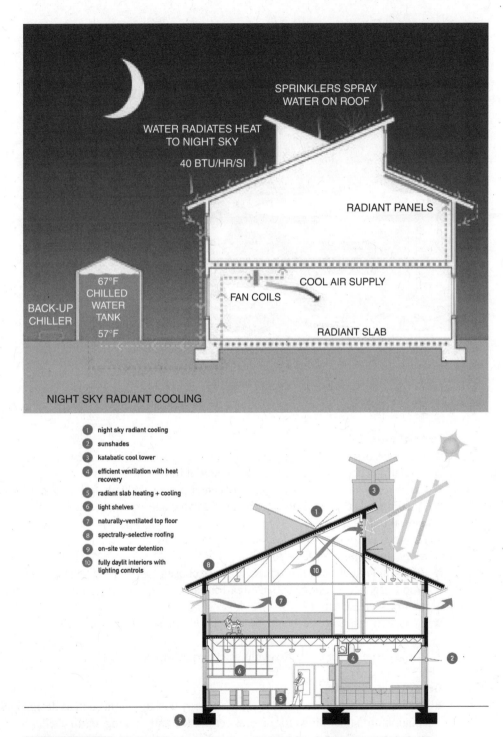

Figure 8.20 Global Ecology Center at night (top), cross-sectional view (bottom), and axonometric view (next page). *(Images courtesy of EHDD Architecture)*

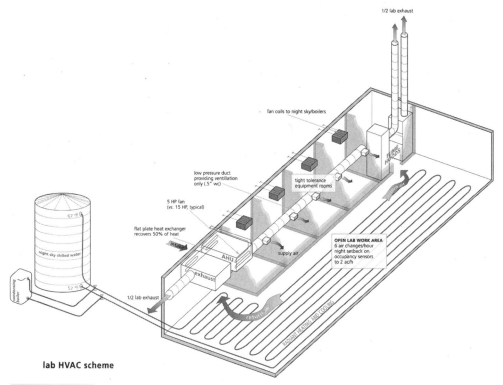

lab HVAC scheme

Figure 8.20   *(Continued)*

Green materials

- Waterless urinals and dual-flush toilets were installed in all the buildings.
- Insulation containing high-recycled material blocks heat gain and loss.
- Green roofs on the library and student center provide habitats for indigenous species.
- Buildings are clustered on an existing ridgeline and stepped to correspond with the natural topography.
- Reduced urban heat-island effect was provided by covered parking, planting green roofs, and using high-albedo paving.
- Paints and other interior products all were chosen for their low chemical emissions.
- Wood products without added urea-formaldehyde, which pollutes the indoor air quality, were selected.
- Cypress trees removed from the site before construction were reused as benches and decks.
- Carpet from factories that recycle used carpet was specified.

Green technologies

- Several measures were taken to reduce energy consumption: such as design and orientation of the buildings for natural ventilation and lighting, earth sheltering, high-performance glazing, and above-regulation insulation.
- The project uses 32 percent less energy than the minimum California's codes mandate.
- Efficient lighting and occupancy sensors reduce lighting costs.

CROSS SECTION

1 PHOTO VOLTAIC PANELS 2 NATURAL VENTILATION 3 RADIANT HEATING 4 SHELTERED PARKING 5 STUDENT CENTER BEYOND 6 ARROYO 7 LIVING ROOF 8 TURBINE VENT. 9 CEILING FAN 10 BERM

Figure 8.21   **Nueva Hillside Learning Complex.** *(Image courtesy of Leddy Maytum Stacy Architects)*

- A 30-kW photovoltaic system provides 24 percent of energy needs.
- High-efficiency boilers and radiant hydronic heating (as opposed to common convection heating methods) are used.

POCONO ENVIRONMENTAL EDUCATION AND VISITOR CENTER

- Building type: Commercial
- Location: Dingman's Ferry, Pennsylvania
- Architect: Bohlin Cywinski Jackson
- Completion date: 2005

The Education and Visitor Center is a flexible gathering space for meetings and environmental learning activities. The organization's mission is teaching environmental responsibility. Visitors pass through a variety of ecosystems (forest, wetlands, etc.) honoring the many aspects of nature, before approaching the building. The main room is on the south and is filled with abundant light and views.

Green materials
- The entrance is clad in shingles made from reclaimed tires collected locally.
- Recycled plastic composite lumber replaces preservative-treated wood.
- Southern windows use solar gain to lower winter heating costs.
- Thirty percent of the cement used was replaced with fly ash, which turns waste into a useful product while reducing CO_2 emissions.

- Using concrete floor slabs that will never require refinishing improves air quality and uses thermal mass to help control heating and cooling loads.
- Light-colored exterior walls and roofs keep surfaces from absorbing large amounts of heat, thus saving energy from reduced cooling loads within the building.
- All materials were chosen with low maintenance in mind.

Green technologies

- Operable windows are used for natural ventilation. Low operable windows on the east and west façades draw in cool air from shaded porches in summer. High operable windows on the south façade exhaust warm air during the summer.
- The steel structure can be mostly disassembled if the future dictates a larger facility.
- Daylight is offered in abundance by the three façades covered in glazing which sustains the lifetime of the artificial lighting system and saves energy.
- The shape of the building mass minimizes the impact of winter winds, which sustains the wall systems for a longer time by not competing with the elements.
- Large overhangs assist in energy saving by better controlling the impact of the high summer sun and the low winter sun for better heat loss or gain.
- An open floor plan is an efficient way of accommodating multiple uses for one space.
- More than one HVAC system services different spaces and is more efficient because it wastes less energy.

SIDWELL FRIENDS MIDDLE SCHOOL

- Building type: K-12 Education
- Location: Washington, D.C.
- Architect: KieranTimberlake Associates, LLP
- Completion date: 2006

Sidwell Friends is a pre-K through 12th grade Quaker independent school. The 55-year-old building was transformed and enlarged by 39,000 ft^2, with the overarching vision of being a model for teaching environmental responsibility. The project is a good example of sustainable construction practices; it is effective, economical, durable, and exhibits a good reuse and maintenance of existing structures.

Green materials

- Solar-ventilation chimneys, operable windows, and ceiling fans help decrease use of a mechanical cooling system.
- An array of materials is designed to take advantage of passive solar design: photo-sensors, occupancy sensors, and, where needed, high-efficiency electric lighting.
- Reclaimed materials include: cladding, flooring, and landscaping stone.
- Interior finishes contain recycled content and low chemical emissions, and are selected from rapidly renewable choices.

- A green roof reduces runoff, and accompanying constructed wetlands treat wastewater for toilets and cooling towers.
- Using native grasses and shrubs improves the quality of infiltrated runoff and reduces municipal water use.

Green technologies
- Water-efficient fixtures allow for the reduction of water consumption as well as reuse of water in toilet bowls and cooling towers.
- A lightshelf, which is part of the facade, allows daylight to penetrate deep into the building.
- Solar ventilation chimneys and a photovoltaic array that generates about 5 percent of the needed electricity further increase efficiency.
- The building should see a 60 percent energy cut in comparison with similar structures.
- Time and effort was expended on many shading and screening devices, along with mechanical ventilation systems.
- A separate utility plant for this building alone was rejected in favor of a central plant for the whole campus.
- Although parking in an underground garage is available, pedestrian transportation is strongly encouraged by bicycle storage, showers, and siting the building near subway and bus stops.

FOSSIL RIDGE HIGH SCHOOL

- Building type: Educational/Public
- Location: Fort Collins, Colorado
- Architect: RB+B Architects
- Completion date: 2005

Fossil Ridge High School's vision was a healthy, comfortable, flexible building, but they also wanted an example of environmental stewardship. Therefore the school board members believed that "building a LEED certified school was the right thing to do." Fossil Ridge became the first Colorado school to earn LEED certification.

Green materials
- Seventeen percent of all materials needed were recycled content, which not only helps reduce waste but promotes reuse of materials.
- Half of the materials were locally manufactured, cutting down on pollution while funding local economy.
- Natural lighting was of utmost importance. Windows were placed on multiple sides of the classrooms, and roof monitors and Solartubes bring light deep into the interior.
- To achieve healthy indoor air quality the building was equipped with operable windows, CO_2 sensors, and paints and finishings with low VOCs.
- Heat wheels are used for heat recovery.
- PVC panels contribute energy.
- Artificial turf on sports fields conserves water that would have been used for irrigation.

Green technologies

- Fossil Ridge uses an off-site energy provider powered by wind turbines.
- Light fixtures are high-efficiency occupancy sensors.
- HVAC coils are connected to occupancy.
- Ice made during the night cools the building in the daytime.
- A raw waterpond is used for irrigation.
- Low-flow faucets and toilets have been installed throughout.
- Seventy five percent of the construction debris was recycled.
- The building is 60 percent more energy efficient than comparable buildings and saves $11,500 annually by conserving water.

CalPERS HEADQUARTERS COMPLEX

- Building type: Commercial
- Location: Sacramento, California
- Architect: Pickard Chilton Architects via Kendall/Heaton Associates
- Completion date: 2005

The CalPERS (California Public Employees' Retirement System) Headquarters Complex is a solar green building designed for the largest public pension fund in the nation. Covering two blocks in downtown Sacramento, the building's program includes 550,000 ft^2 of office space, 25,000 ft^2 of retail space, and parking for 1000 cars—more than 1,000,000 ft^2 in all. Though wanting to project an image of permanence, much concern was given to ecologically sound practices as well.

Green materials

- Steel has been extensively utilized, due to its durability, availability, low cost, and low maintenance.
- Furniture with low chemical emissions was selected.
- Carpet is recyclable.
- A multistory glass atrium contributes natural daylight.
- The ballasted concrete roof is economical and protective, and allows good insulation.
- American cherry wood in the interior has been procured locally, which means low-embodied energy, and has been certified by the Forest Stewardship Council (FSC).
- External shading devices block sun and heat.
- An underfloor HVAC unit improves cooling energy.

Green technologies

- The decision to break the project into two U-shaped buildings which faced each other allowed more daylight to reach interior spaces and permitted an interior courtyard.
- A roof platform shades mechanical equipment and supports a photovoltaic array.
- The use of interior and exterior light shelves uses natural light to provide shade to the lower parts of the space while allowing light to enter the top of the space to create natural lighting for inhabitants. Light shelves also reduce the glare and radiation of the sun.

Figure 8.22   **CalPERS Headquarters Complex.** *(Images courtesy of Peter Aaron/ Esto (top three) and Pickard Chilton (bottom))*

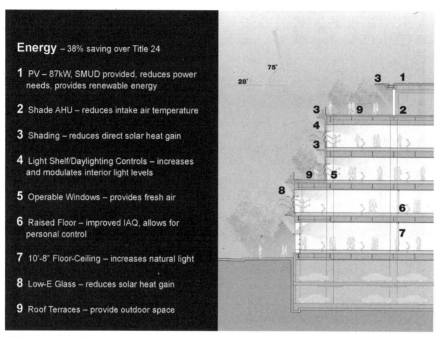

Energy – 38% saving over Title 24

1 PV – 87kW, SMUD provided, reduces power needs, provides renewable energy

2 Shade AHU – reduces intake air temperature

3 Shading – reduces direct solar heat gain

4 Light Shelf/Daylighting Controls – increases and modulates interior light levels

5 Operable Windows – provides fresh air

6 Raised Floor – improved IAQ, allows for personal control

7 10'-8" Floor-Ceiling – increases natural light

8 Low-E Glass – reduces solar heat gain

9 Roof Terraces – provide outdoor space

Figure 8.22   (*Continued*)

- Operable windows provide passive heating and cooling and create healthier environments with clean, fresh air.
- By deciding on underground parking facilities, the heat island effect is dramatically reduced.
- The underground parking enabled the architect to create a more complex and appealing program, with tall atriums and lots of high ceilings and large windows.
- Energy is reduced by 38 percent by the use of systems such as photovoltaic panels.
- Water is heated with recovered waste heat.

CARTER BURTON YOGA STUDIO

- Building type: Residential/Commercial
- Location: Clark County, Virginia
- Architect: Carter + Burton Architecture
- Completion date: 2007

This yoga studio and guest house is small (600 ft^2), but it is crammed only with green technologies. The owners are Buddhists and modernists. The diminutiveness and open, clean space appealed to them. The site rests behind a craggy stone ridge on a 5-acre lot, one hour west of Washington, DC. Native plants remain intact and feature Mountain Laurel, hardwoods, and over one acre of wild blueberries. The program goal was to provide a low-maintenance, energy-efficient structure using experimental materials and techniques that blends with the site without being sentimental. The limited space would require an open plan with upper level storage and built-in storage where possible. Beds and storage cubbies are built into the floor with retractable floor doors allowing for usable floor space during the day and retreat space at night. Architectural and in-house interior design teams worked together to create a cohesive design approach that meets the owners' modern and ecofriendly sensibilities. The shape was inspired by the need for maximum spaciousness within the 600 ft^2, squeezing the bathroom entrance and laundry closet area to its smallest proportion, leaving the largest space for living. The curves provided this space organically while relating to the views and providing passive solar strategies. The details, materials, furniture, and nature are to provide the only art expression. The design firm provided furniture selection and design of custom carbon fiber stools built for this project.

Green materials
- Geothermal heating is efficient in heating, cooling, and providing hot water requirements. The system includes a geoexchanger, a series of tubing runs, and both a liquid-to-liquid heat pump and a liquid-to-air heat pump.
- No VOC was allowed in carpets and paint to eliminate emissions of harmful chemicals and materials into the air.
- Nonpollutant cement also keeps harmful chemicals out of the environment.
- Locally built cabinets came from only 10 miles away.

Green technologies
- Passive solar panels provide energy for the house.
- A green roof composed of indigenous plantings saves 30 to 40 percent on energy costs and retains 70 percent of rainwater, as well as cuts down on runoff.
- Energy efficient LED lighting saves on energy.
- A heat recovery ventilator was installed for fresh air ventilation. The drier air produced inhibits the growth of mold.

CO2 SAVER HOUSE
- Building type: Residential
- Location: Pszczyna, Poland
- Architect: Piotr Kuczia
- Completion date: 2008

The CO2 Saver House was built by Piotr Kuczia on Laka Laka. It is called the "Chameleon House," for, as Kuczia says, "the home blends with its surrounding area." This home uses passive solar techniques for its energy source. It has two green roofs to allow better insulation and for the cleansing of air. The wood used for the fence is unpainted except for a few random planks that sport golds and oranges, which complement the landscape.

Green materials
- Local materials were used to construct the dwelling, like the untreated larch wood that makes up the ground floor exterior and the fence around the home. The reduction of energy spent in transport helps make the house a CO2 saver.

Figure 8.23   Carter Burton Yoga Studio. *(Images courtesy of Daniel Afzal (top and center) and Leesa Mayfield (bottom))*

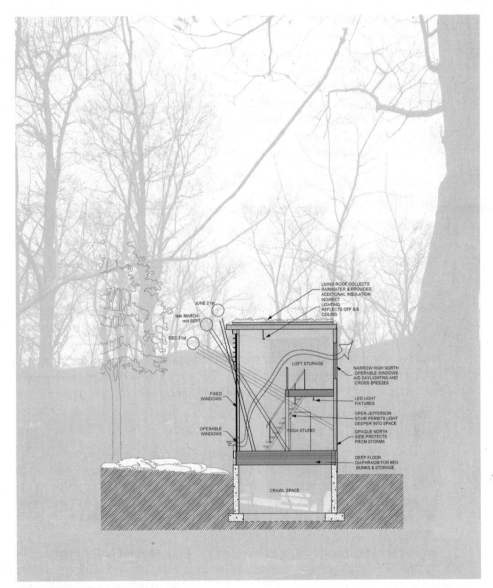

LIVING ROOF COLLECTS
RAINWATER & PROVIDES
ADDITIONAL INSULATION
INDIRECT
LIGHTING
REFLECTS OFF S/S
CEILING

JUNE 21st
late MARCH-
mid SEPT
DEC 21st

LOFT STORAGE

NARROW HIGH NORTH
OPERABLE WINDOWS
AID DAYLIGHTING AND
CROSS BREEZES

FIXED
WINDOWS

LED LIGHT
FIXTURES

OPEN JEFFERSON
STAIR PERMITS LIGHT
DEEPER INTO SPACE

OPERABLE
WINDOWS

YOGA STUDIO

OPAQUE NORTH
SIDE PROTECTS
FROM STORMS

DEEP FLOOR
DIAPHRAGM FOR BED
BUNKS & STORAGE

CRAWL SPACE

Figure 8.23   *(Continued)*

- The black fiber cement panels help optimize solar energy gain. The windows also have fibre cement cladding.
- There are a few hi-tech items that combine with the low-tech building, namely the intelligent building control system.
- The set-in patio with glazing aids in solar energy.
- The ventilation plant has a thermal recovery system.

Green materials

- Some passive solar techniques used are the orientation of the building, 80 percent of the total building faces the south, and the use of thermal mass materials, such as the fiber cement, which covers the central part of the structure, and the concrete floor of the interior.
- The green roofs allow for filtration of the polluted air in the area. Green roofs also act as an isolator for the home, allowing it to hold onto its energy for longer periods of time.
- The home uses only 10 percent of the energy used by an average residence in Poland.

EARTHSHIP

- Building type: Residential
- Location: Taos, New Mexico
- Architect: Anonymous
- Completion date: Ongoing

Taos, New Mexico, is home to what has been called "Earthship Landing Zone." It is a subdivision of a type of alternative housing whose outer walls are partially made up of car tires packed with dirt. All walls except the southern are bermed into the earth, and large windows on the south open to the outside. The houses are so warm and tight that a natural ventilation system is required to keep them from becoming too hot. They are completely self-sustaining and self-sufficient.

Green materials

- Tires which would otherwise be disposed of create the wonderfully thick walls.
- Glass from colorful bottles is used to create mosaics and allow light into the rooms.
- Not only are the materials used readily available, many of them are trash. For instance, interior walls can be made of recycled cans joined by concrete.
- Walls on the inside are often plastered with stucco.
- The tile flooring is made of natural stone.
- The tires are encased in adobe and create thick thermal walls.
- Shaded windows on the inside control solar gain.

Green technologies

- Since the buildings are sunken, the green roof extends down to the ground. The roofs slow stormwater and help prevent flooding.
- The gray water system is completely comprehensive and collects, purifies, then recycles all water used within the Earthship. Gray water is reused several times before finally being flushed.
- Blackwater is filtered through natural clays and gravel before it is used as an irrigation source for the outdoor gardens.
- The southern wall allows light to enter and permits passive solar heating.

■ The occupants have zero utility bills due to the self-sufficiency of the houses.
■ Many designs are horseshoe-shaped to permit maximum light and heat to enter during winter months.

FETZER WINERY ADMINISTRATION BUILDING

■ Building type: Commercial
■ Location: Hopland, California
■ Architect: Valley Architects
■ Completion date: 1996

Fetzer Winery is about more than winemaking; their mission statement begins: "We are an environmentally and socially conscious grower, producer, and marketer of wines." For instance, their acreage is 100 percent organic, they have eliminated the lead capsules that cover corks, and they donate food to food pantries from their experimental organic garden. Early on, Fetzer Winery was one of the first to go green by using only natural sources of energy and cutting back on their waste. Currently they have reduced waste to about 6 to 7 percent.

Valley Architects built a small administration building for the Fetzer Winery in 1996. This building is 10,000 ft² and one of the first of its kind. Valley Architects made the choice to create an entire building of rammed earth and also employed the use of recycled doors and lumber. Photovoltaic panels, added to the building in 1999, now supply almost 75 percent of the building's electricity and energy.

Green materials
■ Rammed earth that is reinforced with a concrete post and beam support system is used for the exterior of the building.
■ Rooftop photovoltaic panels contribute to energy production.
■ Deep overhangs shade the building from solar gain.
■ Vegetated trellising shades the facility and provides ecofriendly design.

Green technologies
■ Water management systems eliminate waste and use gray water to irrigate the vineyards.
■ The winery wastewater is filtered through a natural filtration system that uses gravel, sand filters, and a planted reed bed.
■ A sensor-controlled HVAC.
■ An insulated concrete wall was built to separate cold stabilizing wine from warm fermenting wine, resulting in a power bill reduction of $5000 per month.
■ Night-flush cooling, which means cool air is circulated throughout the night cooling the building before the heat of the day sets in, reduces active cooling during the day.
■ A computerized automated system is used to regulate indoor conditions based on the outdoor temperature and occupancy levels.
■ Extensive daylighting and proper solar orientation is an added design feature.
■ One hundred percent of the energy used is from renewable resources such as solar, wind, and geothermal energy.

PROJECT 7TEN HOUSE

- ■ Building type: Residential
- ■ Location: Venice, CA
- ■ Architect: Schey, Meyer, Gray
- ■ Completion date: 2007

Project 7Ten House is a successful green design, which incorporates green technologies and materials, such as photovoltaic solar panels, rainwater retention, fly ash concrete impediment of radiant heat tubes, heat sensing skylights for cross ventilation, FSC wood cladding, building mass responding to solar orientation and roof slope for rainwater catchment, and shaded southerly glazing to reduce high summer heat gain and promote low winter sun energy.

Green materials

- ■ The new footings used concrete with the addition of 30 percent fly ash.
- ■ The framing of the house was built from FSC lumber.
- ■ Bamboo flooring, as a rapidly renewable material, was used throughout the house.
- ■ Insulation is made from recycled denim, a by-product of blue jean manufacturing that typically ends up in land fills, and is much safer than fiberglass batt insulation.
- ■ Tile and glass in the bathrooms were made from recycled porcelain and glass.

Green technologies

- ■ The demolition of the former home on-site was considerably green; old concrete was sent to a concrete recycling center. Wood not affected by termites was sent to Guadalajara to build low-income housing.
- ■ Energy-efficient windows are valuable both for their insulation properties and their ability to provide natural light.
- ■ Radiant floor heating conserves energy in a number of ways: first, the water is heated by solar panels on the roof, then it is sent throughout the house in a series of tubes. Also, it is a healthier alternative because there is no circulation of dry air.
- ■ The power for the house comes from a series of photovoltaic panels. These panels provide power for the lighting, as well as the radiant heating and air-conditioning that is needed. The panels also double as overhangs, blocking sunlight during the summer months.
- ■ Sensors throughout the house monitor the occupancy and temperature of rooms, turning lights on and off as necessary, as well as controlling mechanically opening skylights to allow the release of hot air.
- ■ Not a single incandescent light bulb was used in the house; rather lighting is entirely compact fluorescent bulbs or LED lighting.
- ■ The water usage of the house is particularly well engineered. A drip irrigation system is used in place of a conventional sprinkler system, which loses water to evaporation. A gray water system was installed to filter and reuse water from the washing machine and showers. Also, an underground cistern collects 75 percent of rainwater from the storm drainage system on the house. Lastly, the landscape surrounding the house

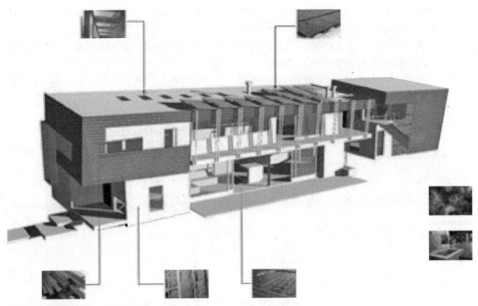

Figure 8.24   **7TEN House.** *(Image courtesy of Melinda Gray)*

is native vegetation along with permeable surfaces, so as to allow water to seep through and back into the earth.

AVAX HEADQUARTERS BUILDING

- Building type: Commercial
- Location: Athens, Greece
- Architect: Alexandros Tombazis
- Completion date: 1998

The Avax Headquarters Building is the hub of one of the major construction companies in Greece. The front of the building is dominated by five white concrete columns which support sunshades, or vertical glass panels. The sunshades are controlled by a central system and rotate like fins, according to light and temperature. The building was designed to "exploit natural lighting and shading technologies." Three basement levels, a ground floor, and five levels above that are fitted into the narrow building. The structure illustrates the applications of "bioclimatic design principles," which refers to the design of buildings based on local climate. Bioclimatic buildings utilize environmental factors like climate, sun, air, water, soil, etc.

Green materials
- The revolving sunshades enable the sun's rays to be controlled and also optimize natural light.
- Horizontal grills complement the sunshade device.

- A steel frame with a mix of reinforced concrete and coffered panels make up the general structure of the building.
- Offices have a granite floor layer over a false floor that gives access to services, such as cooling systems.
- Helical ceiling fans are manually controlled and help control temperature.

Green technologies
- Solar collector panels supply most of the hot water requirements.
- There are "ice buckets" in the underground floor that store cooling energy out of peak times of the day.
- Total energy consumption is approximately 50 percent of that of a conventional building.
- Night air ventilation naturally cools the building, further reducing cooling loads.
- Photovoltaic panels generate some electricity.
- The narrow structure has the advantage of allowing natural lighting to reach the entire depth of the building.

BLUE RIDGE PARKWAY VISITOR CENTER

- Building type: Assembly/Educational
- Location: Asheville, North Carolina
- Architect: Lord, Aeck, & Sargent
- Completion date: 2007

The Blue Ridge Parkway Visitor Center welcomes travelers with multimedia information about things to do and places to visit while traversing the Parkway. The heritage and history of this portion of North Carolina is seen in exhibits, interactive displays, and listening stations. A 22-ft interactive map on the "I-Wall" displays the entire Parkway.

The design firm states that this high-performance, ecological facility's optimized passive solar system, coupled with extensive daylighting and high-efficiency mechanical system with energy recovery, are projected to reduce energy use by 75 percent.

Green materials
- Recycled content in major building components (concrete, steel, and curtain walls) reduces building's energy footprint.
- Regionally harvested southern yellow pine roof.
- Low VOC-emitting finishes.
- Porous modular plaza paving allows stormwater infiltration.
- Vegetated swales treat site runoff.
- Native landscaping reduces irrigation and maintenance demands while providing habitat.
- Existing mature vegetation protected to preserve adjacent Blue Ridge Parkway view corridors.

Green technologies

- The center contains 13 Trombe walls, facing the south, which allow for the maximum benefit of sunlight within the year. During the winter the reverse process takes place, as the trapped air becomes cold from the thermal mass of the concrete wall.
- A CFE computer model studies how the air flow and heat transfer is working and evaluates the efficiency of energy performance.
- The center is also heated with hydronic radiant heated flooring and a high-efficiency HVAC that combines with an energy recovery wheel.
- In the winter, the energy recovery wheel removes the heat from the air that is being exhausted out of the building and transfers it to the air entering the building. In the summer the unit removes humidity from the air before it enters the building.
- A green roof of 10,000 ft^2, planted with drought-resistant plants, cooperates in reducing heat transfer through the roof.
- Photo and occupancy sensors regulate electric lights by turning them off when there is sufficient daylight or when the rooms are empty.
- Ceiling fans in the high ceilings circulate the air from the top of the room back down to the bottom.
- Specifically oriented overhangs and shading devices allow sunlight into the building during the winter season and provide shading when necessary during the summer.
- Operable windows allow for natural ventilation.

PASADENA ECO HOUSE

- Building type: Residential
- Location: Pasadena, California
- Architect: Studio RMA
- Completion date: 2009

The Eco House is a private residential home that aspires to be the first concrete LEED platinum residential building in the United States. This high-performance, active solar house is a good case study combining new technology with modernist design. The floor plan shows a huge, irregular shape with plenty of open space. The house utilized passive and active solar components, green materials, and technologies, including recycled materials, structural concrete insulated panels (SCIPs), fly ash, photovoltaic, and high performance insulation systems.

Green materials

- Sixty percent of the materials are recycled.
- SCIPs are: "prefabricated foam panels with robotically welded mesh on each side and a 3-D truss system welded through the center foam panel." Strength is maximized by the application of a concrete skin to both sides of the panels. This coating also provides fire resistance and good insulation.
- SCIPs, made entirely of recycled materials, will fashion the floors, roof, and walls of the house.
- Cement containing fly ash are used.

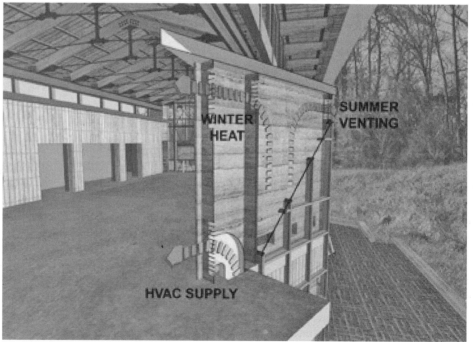

Figure 8.25   Blue Ridge Parkway Visitor Center. *(Images courtesy of Jonathan Hillyer (top) and Lord, Aeck, & Sargent (bottom))*

- Australian koa wood is used, which is sustainable since it is not depleted faster than it can regenerate.
- Recycled glass countertops are made from crushed glass that would otherwise end up in landfills.
- The kitchen cabinets are constructed from FSC certified wood and use nontoxic veneers and laminates.

Green technologies

- The SCIPs panels are connected together by mesh reinforced concrete, further strengthening the structure.
- SCIPs reduce construction time and have far superior insulation. They are 63 percent more energy efficient than a standard 2 × 4 wood frame wall. Use of this technology decreases energy costs and conserves materials.
- Green switch systems are special light fixture switches that conserve electricity throughout a home's lifetime.
- Part of the Eco House's grid dependence is offset by its use of photovoltaics, which will harvest the sun's energy, converting it to electricity.
- An open floor plan increases air flow between parts of the house and decreases heating and cooling costs.
- High-insulation systems on windows and doors further reduce energy costs.
- The appliances that were selected for the Eco House are energy efficient and longer-lasting than standard equivalents.

Modular and Mobile Systems

ABOUT SAVING A PLANET (ASAP) HOUSE

- Building type: Residential
- Location: Sag Harbor, New York
- Architect: Laszlo Kiss
- Completion date: 2008

ASAP is a house "About Saving A Planet." Designed by Laszlo Kiss, a New York architect, it is his answer to the "energy-hogging mansions" he sees around him. It is a prefab, modular design, with 2520 ft^2, and very low in cost (about \$260 per ft^2). It is completely energy efficient and ecofriendly, a "zero-energy house." ASAP is a modernist design, containing mostly open spaces with a few private areas.

Green materials

- Natural renewable materials are used throughout the house. The appliances are low energy consumption.
- Bamboo floors are sustainable and ecosmart.

- An outdoor pool can be heated with solar radiation.
- Pergolas which provide shade also support solar collectors.
- Heating and cooling comes from a ground source heat pump.
- Walls contain R-19 insulation and the roof contains R-38.

Green technologies

- A 10-kW photovoltaic array placed on the roofs powers the house, producing 13,000 kWh a year. Connected to a grid, the photovoltaic will generate more energy than the house needs, thus becoming a power plant of sorts.
- Skylights, pergolas, and shaded decks take advantage of light and reduce solar gain in summer.
- Geothermal heating and cooling produces five units of energy for every unit of electricity used. Savings of up to 70 percent are expected.
- Three prefabricated modules will take only two weeks for the factory to produce and will arrive at the site 80 percent finished.

CARGOTECTURE

- Building type: Residential/Commercial
- Location: Moveable
- Architects: HyBrid
- Completion date: Ongoing (begun in 2003)

Figure 8.26   Three views of the ASAP House. *(Images courtesy of Francine Fleisher)*

Figure 8.26   (*Continued*)

Cargotecture is a word coined by HyBrid in 2003 to describe any building system built entirely or partially from ISO shipping containers. These containers range in size from a one-room office to an entire apartment complex. Because of their modularity, they can be taken apart, relocated, and rebuilt in numerous ways. There are even several roof options including water collection, solar collection, roof deck, or roof trellis. Cargotecture is highly focused on recycling. For example, the container itself is being reused and so are most of the materials inside.

Green materials

- Most of the materials are recycled, including the aluminum windows, sink, and lumber, in addition to the ISO shipping container itself.
- Magnesium calcite paneled walls are non-off-gassing, recyclable, and waterproof. They also take no energy to produce and are a good earth resource.
- LED lights and dimmable low-wattage lighting keep the energy consumption low.
- A fern-based green roof provides a habitat for local species.
- Floor-to-ceiling glazing takes advantage of solar exposure and provides ample amounts of natural daylighting to the interior of the space.
- All appliances have a DC option, so that solar panels can be installed.
- Energy Star appliances and water efficient fixtures are included in all models.

Green technologies

- Removable and adjustable foundation systems allow Cargotecture units to be moved easily, which cuts down on transportation costs and subsequent pollution.
- Water harvesting entails gathering, accumulating, and storing rainwater for later uses.
- Roof water collection irrigates gardens.
- Rain and gray water are conserved and reused.

Figure 8.27   Three views of the Cargotecture. *(Images courtesy of HyBrid)*

Figure 8.27   (*Continued*)

ILLY PUSH-BUTTON HOUSE

- Building type: House, Commercial, Disaster Relief
- Architect: Adam Kalkin
- Debuted at Venice Biennale in June 2007
- Debuted in an exhibit in New York City in December 2007

The Illy Push-Button House is a home designed from a recycled, metal industrial shipping container. At the push of a button it transforms. The container converts into a fully furnished five-room home, opening up like a blooming flower. It represents Illy's dedication to sustainability, art, and certainly, innovation. Computer-controlled hydraulics choreograph the conversion in only 90s.

Green materials

- The container, steel, and nearly every material used are recycled or can be recyclable.
- Bedroom, living room, library, dining area, and kitchen are all made from recycled materials.
- The home is "one continuous moldable surface."

Green technologies

- The house opens by using hydraulic cylinders that are controlled by a computer system inside the kitchen island.
- The whole structure is made of lightweight material in order for the machine to allow the walls to go up and down.
- The house has been on display to demonstrate how mobile mechanics could be used in various situations, such as for disaster relief.

KITHAUS: K1 MODULE

- Building type: Prefab/Residential
- Location: Van Nuys, California
- Architect: KitHAUS Designs
- Completion date: 2006

The KitHAUS Design Company designs prefab modern living spaces using lightweight aluminum for the structure. They are easy to assemble, disassemble, and recycle. The KitHAUS can be used alone or in combination to form larger or smaller modules, depending on the needs of the client. Thus the house can adapt to the present and future spatial needs of the homeowners.

Green materials

- The moisture protection system uses Butyl Rubber adhesive, which will also withstand high temperatures.
- FSC certified wood is used in louvers and canopy.
- Finished birch plywood flooring come from a local product, lowering energy from transportation.

Figure 8.28   Views of the Illy Push-Button House. Closed view (top), conversion stage (center), and fully furnished house (bottom). *(Images courtesy of Adam Kalkin)*

Figure 8.28   (*Continued*)

- The patented M.H.S. aluminum structural framing clamp system quickly assembles and disassembles, and parts can be reused.
- SIPs (structural insulated panels) comprise the walls and ceiling.
- Insulation in windows, doors, and walls reduces the rate of heat transfer.
- Low-E glass: All glass is dual glass glazed with a low-E coating for better thermal performance. It reduces the U-factor by suppressing radiative heat flow. The glass reflects a significant amount of radiant heat, which lowers the total heat flow through the window.
- The exterior is corrugated galvalume and aluminum slat wall, also easily assembled and disassembled, and which can later be reused.

Green technologies
- Prefab construction is earth friendly and reduces waste.
- Prefab construction offers energy-saving designs and improves manufacturing and construction efficiencies.
- Prefab homes are constructed indoors without exposure to rain, dirt, or freezing temperatures.
- Prefab is installed in less time, and is efficient.
- Moisture protection system: A self-adhered, roofing underlayment provides economical, easy-to-apply protection from damage caused by wind-driven rain and ice.
- Solar cells that make up the photovoltaic system convert solar energy into direct current electricity, and reduce energy consumption.

PREFAB FRIDAY: THAILAND'S MODULAR GREEN HOME

- Building type: Residential
- Location: Bangkok, Thailand, Baan Lae Suan Fair
- Architect: Site Specific and Butalah Studios
- Completion date: 2008

This modular green home, created for a home and garden expo in Bangkok, was constructed from four preused shipping containers and prefab modules. It incorporates a simple yet effective design and is applauded for its reuse of materials. The architect wanted to take the four R's (reduce, reuse, recycle, and renew) and incorporate them into a home that would meet a Thai family's specific needs.

Green materials

- The shipping container exterior has not been refinished but retains its natural rough texture.
- A multiplicity of glass windows flood the space with natural daylighting, while screens and shades regulate heat and light.
- Wood flooring is just one example of the natural materials that decorate and furnish the house.
- Stainless steel, a durable and economical material, is used for kitchen fixtures.
- Straw bales form fencing around small kitchen gardens.

Green technologies

- Efficient reuse of available materials and simple, prefab construction methods allow for a sustainable building. No high performance technologies are used in this modular home.
- A gray water system supports water conservation.
- At 1000 ft^2, the two-bedroom home limits environmental impact.
- A plot of land for a garden was integrated into the site.

ZEROHOUSE

- Building type: Prototype/Residential
- Location: Anywhere
- Architect: Scott Specht
- Completion date: 2007

Architect Scott Specht has designed a "completely self-sufficient home that generates its own power, collects its own water, processes its own waste, and is 100 percent automatic." Its ease of assembly, durability, and green technologies are simple and efficient.

Green materials

- Triple glazing and low-E heat mirror glass baffle heat gain, even though windows are large.
- Highly insulated exterior doors help to keep temperatures even.

- Tubular steel frame components comprise the structure. This steel is so durable even winds of 140 mph cannot topple it.
- LED lighting is built into both ceilings and walls.
- A helical-anchor foundation requires no excavating.

Green technologies

- High-efficiency solar panels power the house. A week's worth of power can be stored by a battery and used during sunless days.
- A rainwater collection system processes waste and water.
- All organic waste is processed in a digester unit that converts waste into clean, dry compost so it can then be reused as fertilizer.
- All fixtures are gravity forced, which saves energy because it doesn't require mechanical pumps.
- A high-efficiency HVAC system is only used when needed, thus is more efficient and saves energy.
- All functions of the house are monitored by an array of sensors and regulated by a "house brain" that can be controlled through any laptop computer.
- The home can be erected almost anywhere in less than 24 h.
- The durability of the house allows setup in difficult locations, such as in water or on the slope of a hill.
- The structure can be easily disassembled and rebuilt, or even reused for another project.
- Minimum site disturbance is accomplished because the foundations are anchored to the ground in four points.

LOT-EK MDU

- Building type: Temporary Residential
- Location: Anywhere
- Architect: Lot-Ek; Tolla and Lignano
- Completion date: On market since 2002

The Lot-Ek MDU is a recycled industrial shipping container now turned into a modular home. Various sizes from 640 to 2560 ft^2 are available. The module comes "fully equipped with built-in electrical, plumbing, HVAC, fully insulated, and furnished."

Green materials

- Use of waste shipping containers is obviously an innovative and ecofriendly reuse of materials.
- The purchaser can choose the Compact version or splurge on the Loft style; just add on another container to expand—either up or out.
- Floor-to-ceiling 8 × 8 ft windows aid in passive solar power and give a light and airy interior.
- Wood floors add a touch of luxury.
- Natural plywood interior surfaces are very cost-effective.

Green technologies

■ The module's strengths are its mobility, scalability, and its flexibility.
■ The simple construction method eliminates the ecohazards of other construction projects.
■ Multiple modules can be stored on tiered steel racks, making a sort of village. In this way the individual containers can be tied into power, water, sewage, etc.

THE MICRO-COMPACT HOME

■ Building type: Residential
■ Location: Munich, Germany
■ Architect: Horden Cherry Lee Architects
■ Completion date: 2005

The Micro-Compact Home (m-ch) is a 2.66-m (nearly 9-ft^2) aluminum cubic pod. The pod is upscale "cube living," as it is endowed with advanced concepts and technologies; for instance, a flat screen plasma TV and concealed hi-fi. Embedded in the pod are communication and energy systems. "We wanted to show how a state-of-the-art technology could be integrated into a light-weight compact dwelling," said Architecht Horden. The pod requires no furniture, for it is all included—in aluminum built-ins.

Green materials

■ The timber frame covered with aluminum cladding is lightweight and easily delivered, even with a helicopter.
■ Polyurethane is used to insulate the module.
■ Aluminum frame windows are double framed and double glazed.

Green technologies

■ LED lighting, photovoltaic solar panels, and a vertical-axis wind generator produce energy.
■ Warm air heating controls the temperature.
■ Some of these units can be purchased with the intent to add on. These have strong internal frames so that additional cubes may be stacked on top.

FUTURE SHACK

■ Building type: Temporary Residential
■ Location: Anywhere
■ Architect: Sean Godsell
■ Completion date: 2007

The house is versatile and has multiple uses: postdisaster, temporary housing, third world housing, and remote housing. Once on-site, this low-cost residential prototype is very easy to set up and in less than a day. All the essentials of life are provided: water tanks, solar power cells, compact kitchen and bathroom, fold-out tables and beds, a satellite receiver, food, clothing, and blankets.

Green materials

- The structure is a 20 ft^2 shipping container, easily transportable, mass-produced, inexpensive, durable, but not recyclable because modifications were made to the containers (holes in the top for skylights).
- A pair of steel brackets with telescoping legs accommodates uneven terrain and varying slopes.
- A marine-plywood lined interior resists rotting in an environment high in moisture but is 3 times as expensive as traditional plywood.
- The parasol roof shades the container and protects the water supply. An interesting feature of this roof is that it can be personalized. Its galvanized steel frame is designed to accept thatch, mud and stick, or palm fronds, any of which would increase the insulation R-value.
- A roof ladder provides egress to the roof.
- An access ramp folds out to provide access to the interior through a hydraulic door, and acts as a front porch.
- Water tanks support the water requirements.
- The furniture folds out of the walls for multifunctional purposes.

Green technologies

- Solar power cells generate electricity to operate the shack's lighting.
- A satellite receiver provides a connection to the world.
- Natural ventilation is provided by a series of operable panels in the roof.
- Heat is obtained with bottle gas.

POWER PODS

- Building type: Residential/Commercial
- Location: Lawrence, MA
- Architect: PowerHouse Enterprises
- Completion date: 2007

The PowerPod is a self-contained modular unit measuring 480 ft^2. It maximizes space and can be shipped in its entirety on one flatbed trailer. The facade has something of the appearance of a regular house. The cost is approximately $200 per ft^2.

Green materials

- Very high insulation in the walls (SIP R-28) stabilizes the temperature. Advertisements claim the Pod "uses fuel with a teaspoon."
- The solar butterfly roof hides the solar collector and catches rainwater as well.
- All the materials used to create the floors and walls are readily available and can be produced or harvested sustainably.
- Solar electric panels located on the roof generate electricity and heat.

- The radiant-heat flooring is highly effective at absorbing sunlight during the day and radiating that heat at night.
- The kitchen countertop is composed of recycled newspapers, while shelving and sinks are stainless steel.
- Decks can be placed along the building's longest sides.
- The buyer has the option of selecting a dual-flush toilet or a composting toilet.
- A steel pier foundation is sturdy, adjustable, and makes setup uncomplicated.
- The high ceiling eliminates the feeling of being boxed in.
- Numerous windows offer an airy feeling and maximize views.

Green technologies
- The metal roof heats by way of plastic water tubes under the surface. As water in the tubes heats up during the day, the tubes distribute it throughout the house, supplying hot water and heat.
- The angled (butterfly) roof is designed to collect and funnel rainwater so it can be redirected and used in the home or garden. The roof is also designed to "bounce" sun rays from the front angle to the back, diverting heat from the roof. The extended style of the roof forms effective sunshades, reducing solar gain.
- The interiors of the Pods receive enough sunlight directly and indirectly that they do not need to be artificially lit on sunny days, reducing energy consumption.
- Transportation of the homes requires only a flatbed truck.

SPLIT HOUSE MODULAR

- Building type: Residential
- Location: Beijing, China
- Architect: Yung Ho Chang
- Completion date: 2004

The house is divided into two separate rectangles or wings connected by a bridge, while the mountain behind gives a sense of enclosure. It was inspired by Chang's idea of traditional Chinese courtyard dwelling.

Green materials
- Rammed earth walls maintain a stable temperature, summer or winter and are virtually fireproof. The soil, which must be mostly clay is put into molds and compacted, producing a product that is nearly as durable as concrete.
- In the interior laminated wood beams are exposed.
- Floors are stone tile.
- Large glass windows open up the space.
- A stream running through the property flows under the glass floored foyer and through the courtyard.
- A glazed floor composes the bridge connecting the two modules.

Figure 8.29   Two views of the Split House Modular. *(Images courtesy of Asakawa Satosh)*

Green technologies

- The two sections of the house are completely separate, except for the connecting bridge, and can both be inhabited or one can be shut down to save on expenses.
- Since the concept is modular, the house can be positioned as just a single unit, or they can be placed parallel to each other or at right angles. The various positions allow for many central courtyard possibilities as well.

MODULAR-3 HOUSE

- Building type: Residential
- Location: Kansas City, Kansas
- Architect: University of Kansas Students
- Completion date: 2006

Studio 804 is a design/build program at the University of Kansas, headed by Dan Rockhill. Each semester, the students create a modular home from site selection to installation. This is the third student project of this sort, hence the name Modular-3 House. The profits made from selling the previous modular homes are used to construct the next. The design of Modular-3 consists of 6 modules put together on an elevated platform to create a two-bedroom, open-plan home.

Green materials

- The modules were created in a warehouse, then shipped to the site in Kansas. Prefabrication reduces construction time.
- A raised pier foundation enables low-impact implementation.
- A movable storage wall allows modulation of interior space.
- The Douglas fir exterior (a fast-growing softwood) is sealed with an environmentally friendly and biologically safe product.
- Densely packed recycled cellulose insulation (made from recycled newspaper fibers) are used in floor, wall, and ceiling spaces. This insulation has a good fire rating.
- Bamboo flooring, a highly renewable wood source, is featured throughout the interior, and a nontoxic glue was used to install it.
- High recycled-content steel was used on the steel I-beam supports, ramp, and railing.
- High-fly-ash content concrete was used for all concrete work.
- The curtain wall and door are from 50 to 75 percent recycled aluminum.
- Argon-filled Low-E annealed glass forms the bedroom and hallway windows.
- Recycled wooden formwork is used for the concrete staircase that leads up to the house and the retaining wall.
- Crushed rocks wrapped in filter fabric are placed under the downspouts to filter rainwater before permeating into the ground.
- Double cellular shades allow the curtain walls to hold more heat in at night and in the winter.
- Seventy XL solar control low-E glass is used in each curtain wall system.

Green technologies

- The compact modular design cuts down on exposure to the elements, thus reducing heating and cooling needs.
- Passive solar gain is achieved by positioning the main living space to the south. The south-facing glass is pushed back 4.5 ft to create an adequate overhang to shade it from harsh summer sun.
- Large south- and east-facing curtain walls are used to maximize natural light and reduce the dependence on electric light.
- Operable windows are located to allow for cross ventilation.
- A heat pump is used for more efficient energy generation.

SUMMER CONTAINER

- Building type: Residential/Temporary
- Location: Espoo, Alabama
- Architect: Markku Hedman
- Completion date: 2000

The Summer Container is a mobile timber framed container building meant to be picked up and carried away—perhaps on the back of a trailer, or even a large sled! The container uses the basic workings of a matchbox for inspiration, as it folds into itself as a compact box, secure for transportation. It can be occupied by one to two people and is designed for the purpose of mobility and flexibility. Facilities include a kitchen area, a worktable, and a living/sleeping section with storage space. Once settled, windows open up and the interiors slide open to create a small living area.

Green materials

- The exterior is made up of different kinds of plywood boards coated with a laminate.
- Phenolic resin (a polymer made of resin and a base material of paper, glass, or cotton) is the product used to laminate the plywood.
- Insulation is a thicker board of polystyrene (common thermoplastic polymer) tucked between the inner and outer layers of plywood and glued to them for stability.
- The use of timber framing (versus steel, for example) is lightweight and affordable.
- Materials used are more functional than ecofriendly. However, the use of wood instead of carbon fiber, for example, leaves a smaller carbon footprint.
- Since the main objective is to pack up and move easily, low weight materials are a necessity.

Green technologies

- The polystyrene board insulation (50–70 mm thick) is low maintenance and economical.
- As the Summer Container is seen primarily as a holiday getaway home, very little technology is involved. The containers can be outfitted with a kerosene cooker and a sink.
- Electricity is produced by a wind-driven generator or solar panels.

PAPER LOG HOUSES

- Building type: Temporary Residential
- Location: Kobe, Japan
- Architect: Shigeru Ban
- Completion date: 1995

The Paper Log Houses are houses built with walls of paper tubes and foundations of sand-filled beer cases. As a response to the Kobe earthquake of 1995, Architect Shigeru Ban developed a simple temporary home for victims who, six months after the disaster, were still living in tents. The structure was designed for unskilled labor with easily accessible, cheap materials. The total cost was less than $2000, and the homes could quickly be disassembled for recycling. An important green feature of these structures is that they are built from recyclable materials that are sustainable, environmentally friendly, and free of pollutants.

Eight months later Ban built a church using the same techniques. His glimpse of people praying under umbrellas in the mud and freezing rain near their destroyed church was his inspiration. The church is rectangular, built from paper tubes with wooden stoppers at both ends. The surprisingly strong columns formed in this way can support about half as much as a column of wood. The roof is from white teflon-coated fabric and the walls between the tubes are clear plastic louvers. The space near the altar has tubes lined up close together to form a backdrop with storage space behind.

Green materials

- Beer cases were donated by liquor sellers.
- Sand bags or loose sand filled the beer cases to add stability.
- Paper tubes are readily available in different diameters and thicknesses, or they can be constructed easily on the disaster site.
- Self-adhesive waterproof sponge tape is used to connect the paper tubes of the walls and to insulate the interior space.
- Tent-fabric canopies serve as roofs for the structures. The roof canopy can be lifted up to allow for ventilation in the summer.

Green technologies

- Ban is a pioneer in paper tube architecture. He learned how to make load-bearing columns and mold quite beautiful trusses. He also designed a way to fire and waterproof the houses.
- The small interior space can be heated, if necessary, by a small portable heater.
- The paper tube technology is being used in other countries for disaster relief. Alterations were made as different materials and different requirements dictated. For instance, in Turkey, shredded wastepaper was inserted into the tubes. In India, rubble substituted for beer cases as the foundation material.

SUSTAIN MINIHOME

- Building type: Residential
- Location: Anywhere

- Architect: Sustain Design Studio
- Completion date: Ongoing

The Sustain MiniHome comprises a wide variety of modular and mobile home systems that are prefabricated and designed for those that move and want to take their personal homes with them. While most other modulars are stay-at-homes, this one is literally a travel trailer. Think of traveling through America's parks with this beauty. Altius Architecture, Inc., has come up with multiple designs for many different countries, even personally made rooms for small areas like London. These homes are built on their own chassis with wheels, made to move on demand.

Green materials
- It uses one-tenth of the materials and resources to construct as does a 2000 ft² home with a basement.
- Materials are premade and then shipped to the desired location.
- There is no vinyl, no formaldehyde, no toxic adhesives or finishes.
- Everything uses all water-based or plant oil-based finishes. The wood inside is rubbed with beeswax.
- Composting toilets are included.
- No CFCs or HCFCs are found anywhere.
- All woods are FSC certified.
- Natural rubber flooring is available.

Green technologies
- A sterling engine cogenerator can run on biodiesel, which heats the water, gives electricity, and also creates space heating.
- Wind turbines and solar panels (which can be moved on a pivoting system corresponding to different positions of the sun in different seasons) create the electricity for the entire housing unit.
- It uses one-hundredth of the electricity of a 2000 ft² house. The housing units can also be zero-cost energy efficient.
- A roof garden creates passive cooling through evaporation.
- The MiniHome uses one-tenth water and gas.
- Stoves and refrigerators can be run with a small amount of propane.
- There is a high natural ventilation rate when the windows are open and a constant fresh air supply when the windows are closed from the use of heat-recovery ventilators.
- It is durable and requires low maintenance.
- Mechanics are built-in to the flooring of the unit.

SWELL HOUSE

- Building type: Residential
- Location: Venice, California
- Architect: Jennifer Siegel
- Completion date: 2006

Figure 8.30   Three views of the Sustain Minihome.

(Images courtesy of Andy Thomson)

Figure 8.30   (*Continued*)

The Swellhouse is a model of prefab housing design that was planned around earth-friendly technologies. The three bedroom house was constructed in panels that were then put together in a quick and easy construction process. This type of construction not only reduces time and labor but also reduces site waste. One goal of the project was to create a "mutated outgrowth of the existing home...salvaging the original footprint and roof to decrease demolition waste." The house was built on-site and designed on a grid system. Exposed steel columns and high ceilings are just a couple of the prominent features of the project.

Green materials
- A steel frame on the ground floor gives strength to allow for glazing.
- The cladding is made up of interlocking SIPs (structural insulated panels), which add to the overall thermal and acoustic insulation of the home.
- Materials were chosen to reflect a certain budget and lifestyle.
- Large sun-system glass panels contribute to the flow between indoor and outdoor spaces.
- Acrylic plastic bars and aluminum louvers help to make up the glass panels.
- Fiber and cement board panels on the top floor do double duty as rain screens and water barriers while still allowing fresh air to flow between the wall and cladding.
- Ipe decking is both cost-effective and ecofriendly.

Green technologies
■ A panelized system of construction helps to reduce time, labor, and site waste.
■ A radiant heating system helps to reduce energy costs.
■ Prefab design is clean, controlled, and efficient.

WEEHOUSE

■ Building type: Temporary Residential
■ Location: Anywhere
■ Architect: Alchemy Architects
■ Completion date: 2003

WeeHouse is a trademarked name for a contemporary, prefabricated structure originally designed by Geoffrey Warner, principal of Alchemy LLC. The standard
WeeHouse can be used as a home, office, penthouse, and studio. The WeeHouse is prefabricated and then lowered onto the site ready to move in. A one-room space that
looks like a modern cabin, the interior is warm and inviting with plenty of wood and
huge glass expanses (both long ends appear to be almost entirely window) which provide views of the surrounding countryside. Limited to 14×62 ft (350 ft^2) to allow
delivery by truck, they can be stacked for more floor space.

Green materials
■ Low-VOC and green-certified materials are used in the modules.
■ The structure is framed in steel and wood.
■ Ebonized oak ply shelves and fin wall, bed frames, translucent bookcase that sepa-
rates the two beds, stair, and rolling underbed storage were all fabricated in the
architects' shop.
■ The cabin was rough-wired for future service, although site is currently off the grid.
■ The outhouse was fabricated from leftover house materials. A composting toilet
may be inserted under the bunked bed in the future if needed.
■ More green finishes and materials are planned for future models.

Green technologies
■ Energy Kits include: "solar panels, roof mounting hardware, charge controller, battery backup bank, battery interconnects, combiner, breaker, and inverters."
■ The house was fabricated off-site in a warehouse in the dead of winter (Jan/Feb)
for installation on-site 1-3/4 h away. The project was partially driven by the idea
that a house that could arrive on-site 100 percent complete would have significant
advantages over site-built or even partially prefabricated work. A steel frame provided rigidity and an effective means to bolt the prefabricated porch to the main
house on-site.
■ The exterior sheathing is cementitious panel painted with a latex paint imbedded
with iron grit and oxidized to form a natural weathering layer.
■ The main green technologies used are prefabrication off-site construction, along
with small size and efficiency.

Figure 8.31   **Two external views of the WeeHouse.** *(Images courtesy of Alchemy Architects)*

Solar Decathlon Projects

The Solar Decathlon is an architecture/engineering joint design competition, sponsored by the U.S. Department of Energy and the National Renewable Energy Laboratory, to produce the most effective, energy-efficient, solar-powered house that also displays excellent design features and beauty. Taking place every alternate year, the first competition began in 2002, and the fourth will take place in the fall of 2009. Selected teams of university students compete in designing, building, and setting up their creations. Seed-money from the Department of Energy ($100,000 per university) helps fund the projects. The completed structures are displayed on the National Mall in Washington, D.C. for three weeks during the fall. The public is invited to tour the completed homes, learning about the best in solar energy and energy efficiency.

The competition teaches students hands-on ways to learn the principles of good design and efficient-energy production, not by inventing new technologies, but by using commercially available materials that prove a sun-powered home can be produced from today's technologies and materials without sacrificing modern comforts and appearance. Many approaches are seen in the finished entries. Diverse geographic locations are represented, as well as low-income dwellings, urban renewal projects, and mass-producible habitats.

There are 10 categories in which teams can win up to 150 points each, totaling 1000 points. Some categories are worth more points than others; for instance, the architecture category has a total of 100 points, and the net metering category earns 150 points. The 10 categories are:

Architecture: Entries are judged on the design aesthetics of the entry, regardless of its energy performance.

Market Viability: Judges evaluate whether the homes are market ready.

Engineering: Entries must meet mechanical, electrical, and plumbing efficiency values.

Lighting Design: The quality and quantity of both daylighting and electrical lighting are observed.

Communications: The teams are evaluated on how well they present their entry with public presentations and Web sites.

Comfort Zone: The homes are assessed to see how well they maintain a stable temperature and relative humidity.

Hot Water: Each entry must demonstrate that the structure can provide enough energy to heat all the water needed for domestic use.

Appliances: The homes are completely furnished and must provide enough energy to do common household chores, such as cooking, washing dishes, doing laundry, and operating computers.

Home Entertainment: The teams must hold two dinner parties, operate TVs, computers, etc.

Net Metering: The teams are rated on how much surplus energy the house produces. Surplus energy is used to power electric cars that could be used for transportation.

The government views these contests as more than just another student project. The teams are ambassadors to the future—creating energy-efficient, marketable home designs powered by solar energy. The President's Solar America Initiative, which seeks to make solar power match conventional forms of power by 2015, sees the competition as an investment in the future.

For more information, see the Department of Energy Web site http://www.solardecathlon.org

Solar Decathlon 2002

Winner: University of Colorado at Boulder and Denver

AUBURN UNIVERSITY 2002

Auburn University entered a design that is reminiscent of traditional southern buildings but updated with all modern technologies. The team incorporated the once common "dogtrot" or breezeway, which is a central porch that divides the house into two parts. The breezeway allows airflow to enter and pass through the house. Also traditional is the metal roof that overhangs the house to provide a large, shady area. But aside from these traditional touches, the house is completely modern.

Green materials

- The metal roof, common in the south, is made of recycled steel and aluminum. It reflects some of the sun's heat away from the house, reducing energy spent in cooling.
- The windows are triple glass systems including double low-E coating with argon gas that provides exceptional performance throughout all seasons of heating and cooling.
- Flat plate solar collectors use water-based fluid, which is less harmful to the ecosystem, to capture energy from the sun, reducing the amount of energy consumed.
- SIPs make up the floor, exterior walls, and roof surface. Not only are SIPs a strong building system, but the insulation makes them energy efficient and cost-effective.
- Photovoltaic solar cells convert solar energy into direct current electricity.
- Solar megaphones or "light chimneys" are skylights filled with prisms that utilize special optics to direct sunlight into interior spaces, thus reducing the dependence on artificial lighting.
- Water tubes are wide hexagonal prisms that use the principle of thermal mass to capture solar energy in the winter and that absorb excessive interior heat in the summer.

Figure 8.32   Auburn University 2002—Solar Decathlon Project.
(Image courtesy of U.S. Department of Energy (DOE)/National Renewable Energy Laboratory (NREL))

Green technologies

- The south façade is open, with only passive energy, while the north contains the more private space. Positioning the building and interior rooms correctly gets the most from natural lighting and shading.
- Passive solar energy entails capturing solar energy via passive systems integrated into the structure. They rely on temperature gradients created by the absorption of radiation to move energy or fluids.
- Energy-efficient windows and doors reduce energy consumed and help keep a stable temperature.
- A Dogtrot House is a covered open space between two closed spaces, providing a sheltered, breezy area to gather during the summer months. A surround porch provides natural ventilation of excess heat.
- The large, water-filled cylinders placed around the house help heat in winter and cool in summer.

CARNEGIE MELLON 2002

Carnegie Mellon is located in the city of Pittsburgh, which inspired the team to focus on designing a house that could fit on a smaller city lot. The challenge then was to find sufficient space. The students decided to "break the rules" and create a two-story house, which maximizes space on a narrow lot. An added feature is a rooftop deck with a garden.

Green materials

- The rooftop deck garden is located under evacuated-tube solar collectors fitted into a canopy and used for heating water.
- The batt insulation in the house contains unusual recycled material—used blue jeans from a local jeans factory.
- The house was framed in metal, rather than wood.
- No formaldehyde was used in any of the materials.
- SIPs make up the floor, exterior walls, and roof surface. Not only are SIPs a strong building system, but the insulation makes them energy efficient and cost-effective.
- The walls have an insulation value of R-33; the roof has an insulation value of R-50.

Green technologies

- A two-story house was chosen since the team believes ranch style homes take up too much space on the planet.
- The second story provides a loft bedroom along with a green roof terrace.
- Forty two photovoltaic panels produce the electricity.
- Heating and cooling is supplied by a water source heat pump.

Figure 8.33   Carnegie Mellon 2002—Solar Decathlon Project.
(Image courtesy of U.S. Department of Energy (DOE)/National Renewable Energy Laboratory (NREL))

CROWDER COLLEGE 2002

The Crowder College team's home has hints of a French cottage, with a deck or porch surrounding three sides of the building. The design is visually appealing with a wood cabin look. The group was particularly careful to avoid the use of gasoline-driven machinery, even in the offloading and assembly stages. After setup, the home used electricity that came from a trailer-mounted portable solar-electric system.

Green materials
■ Amorphous thin-film photovoltaic modules, rather than crystalline silicon modules, were used. The photovoltaic modules were integrated into the metal roof so that they were nearly invisible.
■ Radiant floor heat was installed.

Green technologies
■ The team addressed their state's problem with "sick building syndrome" by paying careful attention to ventilation.
■ Waste heat from the photovoltaic modules heated the water by means of copper tubes attached at the back, which allowed each module to act like an absorbing plate in a flat-plate solar water heating collector.

Figure 8.34   Crowder College 2002—Solar Decathlon Project.
(Image courtesy of U.S. Department of Energy (DOE)/National Renewable Energy Laboratory (NREL))

TEXAS A&M UNIVERSITY 2002

Texas A&M used cutting-edge construction science to develop its solar-powered home. Though the technologies used were traditional, the students modified and improved them. For instance, the refrigerator appliance was redesigned to make it more efficient. Perhaps the most outstanding thing about the building is its interior thermal wall. This wall is filled with water, which is circulated, heated, and cooled in the wall.

As one team member said, "You don't have to eat only vegetables and live off of the grid. You need to take a couple of these energy efficiency ideas, start with one at a time, and get to a goal of using less energy."

Green materials

- The interior thermal wall with circulating water for heating and cooling is based on refrigeration technology and condenser systems.
- Structural insulated panels (SIPs) formed the entire framework, providing affordable insulation that is durable and effective.
- Pedestal footings elevate the house above the ground, which leaves the earth undisturbed.

Green technologies

- A creative concept called the groHome system, invented by the team leader, uses modules that are dimensionally coordinated to all fit with each other. For instance, there is the groWall, the groFloor, and the groRoof, all of which coordinate and can be altered or replaced with other components if desired. It is "an open source kit of parts that uses green practices."
- The photovoltaic modules (3.6 kW) work with the refrigeration-emulation thermal water wall and the water source heat pump in order to supply the house with optimal energy usage. Technologies emulate a refrigeration technique that could dramatically improve conventional methods of cooling. The option is very economical and efficient.

TUSKEGEE 2002

The Tuskegee team used an open breezeway running north and south through the interior to provide ventilation. They conquered another challenge—of adding a second story or elevated loft. This was difficult because an 18-ft height limitation is mandated in order to prevent one team's house from shading another. The interior of the house feels very spacious and makes it easy to heat and cool, and the southern-style screened porch provides an inviting entrance.

Green materials

- The main material used on the project was wood, a light and effective way to build modular homes. FSC certified woods, such as the plywood siding and the all-wood interiors, were used.

Figure 8.35   Texas A&M University 2002—Solar Decathlon Project. *(Image courtesy of U.S. Department of Energy (DOE)/National Renewable Energy Laboratory (NREL))*

- Batt insulation was used in the floor and walls.
- The roof was insulated with rigid foam insulation.
- Windows are double-paned insulating glass.
- Compact fluorescent lamps are used as being more efficient and generating less heat than incandescent lamps.
- Lighting is computer-monitored and adjusts depending on how much natural light is available.
- Light-colored paint on interior walls aids daylighting.

Green technologies

- Forty BP solar photovoltaic modules helped keep the house very energy efficient. It is projected to provide enough power for nearly six sunless days.
- Hot water is obtained by a thermal panel with underlying pipes. A backup electric water heater is there in case of need.
- A high-efficiency heat pump is used for the heating and cooling of the house.
- The bedroom in the loft is connected by a dogtrot to an adjacent screened porch on the west. This design allows the upstairs sleeping chamber, which would otherwise be unbearably hot and stuffy, to be filled with cool breezes.
- Large windows on the south use passive solar heating to warm most of the house. If necessary, a central heat pump unit can be used.

Figure 8.36   Tuskegee 2002—Solar Decathlon Project. *(Image courtesy of U.S. Department of Energy (DOE)/National Renewable Energy Laboratory (NREL))*

■ The plywood siding is designed to release solar heat before it has a chance to radiate into the interior.

UNIVERSITY OF NORTH CAROLINA AT CHARLOTTE 2002

Students at the University of North Carolina at Charlotte had an extremely innovative idea. They wanted appliances that would reduce the energy load on the house, and they found them—on a yacht. All the compact, efficient, 120-V appliances were purchased from a marine supplier.

Green materials
■ The house uses "Kalwalls," which are insulating panels that keep out heat from the sun while allowing light in.
■ SIPs were the primary material used in construction.
■ The walls have an insulation value of R-19, while the roof is R-40.

Green technologies
■ Passive solar provides much of the heating.
■ Space cooling is achieved with a water source heat pump along with natural ventilation.

Figure 8.37   **University of North Carolina at Charlotte 2002—Solar Decathlon Project.** *(Image courtesy of U.S. Department of Energy (DOE)/National Renewable Energy Laboratory (NREL))*

- Water is heated with a flat plate collector and a 15-ton water source heat pump.
- The house has a 140-gal water storage tank.

UNIVERSITY OF COLORADO AT DENVER AND BOULDER 2002

The main focus of the University of Colorado's team was to prove that solar energy could work in a traditional home. Their design looks like a typical house you might see on any small-town street. Their entry was, as one of the team members put it, "a beautiful house that also happened to be highly efficient and solar-powered." They wanted people to like the look of their house, and its positioning on a corner lot with a welcoming front walkway achieved that goal. Their creative and innovative use of prefabricated modules (explained below) sets a new precedent for efficient building.

Green materials
- All materials are mass-produced and easily available.
- High-performance windows prevent thermal gain.
- Biodegradable, recycled, and rapidly renewable materials were chosen.

Green technologies
- The team used their own BASE+ concept (Building A Sustainable Environment concept). In brief, it is an "adaptable construction method for repeatable, site-specific housing" which uses three types of prefab modules (base, spec, and utility)

that can be joined and arranged in many different ways to match specific site, budget, or climate requirements.

- The basic module is constructed from SIPs. In the team's entry, one basic module makes up the bed/bath area and another basic module is the office/living area.
- The spec module can be customized and personalized to match an individual client or a specific site. In this project, the spec module houses the kitchen/entry and is personalized with frame construction as well as the SIPs. It also features more windows than the base module.
- The utility module is the energy and equipment center where all technical equipment is housed. In the Decathlon project, the utility module (or TechPod) has a metal frame that plugs into the other modules.
- The three modules are joined by a central space.
- Photovoltaic panels on the roof supply energy; surplus energy, if generated, is stored in batteries in the utility module.
- Tube solar collectors were used to heat water.
- Shading devices include fixed overhangs, vertical fins, and sliding barn doors.
- A passive solar system orients the house on an east-west axis to maximize north-south daylighting.

Figure 8.38 **University of Colorado, Boulder and Denver 2002—Solar Decathlon Project.** *(Image courtesy of U.S. Department of Energy (DOE)/National Renewable Energy Laboratory (NREL))*

UNIVERSITY OF DELAWARE 2002

The interesting shape of the University of Delaware house looks something like the capital letter D with the rounded portion of the letter facing south. The curved wall is mostly windows, which considerably enhances the appearance. Inside, the circular open space is exciting and functional. The students chose not to construct a typical boxy structure, even though they knew their unusual design might have energy-related risks. Their high score in the energy portion of the contest affirmed the decision.

Green materials
- Prefabricated panels of solid expanded polystyrene reinforced with steel struts compose the roof and walls.
- The exterior is coated with a polyether-based elastomeric coating.
- Flooring is of a composite material containing a fiber-reinforced polymer network filled with polyurethane foam.
- The radiant floor heating system uses Warmboard panels made of fluid tubing fitted into a plywood underlayment and aluminum sheeting. Over this goes nearly any type of flooring desired.
- The windows are titanium coated with an insulation value nearly 3 times that of a traditional window.
- Forty photovoltaic panels supply energy to heat and cool the house. A battery system can store enough energy for three days of reserve power.

Green technologies
- An underground heat pump pulls unwanted heat into the cool ground. In the winter, when the ground is warmer than the air, the system works in reverse.
- The photovoltaic panels were inclined 20° to the horizontal to achieve maximum exposure to the sun.
- The house faces the south, utilizing the roof's slope to increase energy production.
- The front of the house is in the shape of a semicircle (chosen for aesthetic purposes) with large vertical windows that provide ventilation and daylighting.
- Large windows on the east allow morning sun to warm the interior and thermal floor mass.
- The western facade is designed with smaller windows to decrease unwanted thermal gain from the lower position of the setting sun.
- On the south and north, windows allow passive ventilation.
- A double door system at the entrance acts somewhat like an airlock in that the space between the doors forms a separation between the controlled temperature of the inside and the weather outside.
- The rounded front of the house allows more photovoltaic panels to be fitted on the roof.

UNIVERSITY OF MARYLAND 2002

The University of Maryland's home does not appear to be a solar-powered, energy-efficient house. And that's all right with the team, since that was the focus of their design. They hoped their entry would look very much like any other traditional residence. The bay

Figure 8.39   University of Delaware 2002—Solar Decathlon Project.
(Image courtesy of U.S. Department of Energy (DOE)/National Renewable Energy Laboratory (NREL))

window and skylight are nice home-town touches, while the generous deck on the north gives the house added presence.

Green materials
- SIPs (structural insulated panels) were found to be the best option for the frame.
- The SIPs gave an R-35 insulation rating to the walls and an R-40 rating for the roof.

Green technologies
- Ninety six photovoltaic modules arranged on the back roof gather the energy for the house.
- The hot water system was superior in that it supplied hot water for both the domestic use and also for the radiant floor heating system. The process uses evacuated glass tubes containing a collector medium which transfers heat. In the winter, this system is preferred as it collects more heat.
- A photosensor lighting system detects the amount of light in a space and regulates how bright or dim the lights should be.
- Windows on the south and the north side provide optimum daylighting without a significant loss in the energy needed to keep temperatures stable.
- The appliances selected were energy efficient.
- The team had time to pour a concrete slab floor, which in turn gave them the opportunity to install radiant heating.

Figure 8.40   University of Maryland 2002—Solar Decathlon Project.
(Image courtesy of U.S. Department of Energy (DOE)/National Renewable Energy Laboratory (NREL))

UNIVERSITY OF MISSOURI-ROLLA AND ROLLA TECHNICAL INSTITUTE 2002

The University of Missouri-Rolla and Rolla Technical Institute team chose a ranch style design that would be seen as familiar and comfortable, rather than futuristic. "We didn't want it to look like a high-tech science fair project," said their faculty advisor. The sunroom on the south not only contains all the controls, but is perhaps the most outstanding feature of the home.

Green materials

- Steel studs were chosen as the construction material, for their strength, recyclability, and resiliency.
- Extruded polystyrene foam insulation is used on the walls and floors (R-21) and ceiling (R-40).
- Wood paneling gave the house a cozy, comfortable appearance.
- Since some members of the team were enrolled in RTI, a vocational school in Rolla, the house had the advantage of unique furnishings such as cabinets, an innovative bookshelf, and a deck.
- A 40-gal storage tank holds hot water.

Figure 8.41   **University of Missouri-Rolla and Rolla Technical Institute 2002—Solar Decathlon Project.** *(Image courtesy of U.S. Department of Energy (DOE)/National Renewable Energy Laboratory (NREL))*

Green technologies

- The house is composed of three modules which were joined together on-site.
- Multicrystalline photovoltaic cells are used rather than monocrystalline. The team considered the uncertain Missouri weather and decided that multicrystalline panels would be the better choice.
- A sunroom on the south side uses high performance floor tiles that absorb heat during the day and emit heat at night. This is a high-tech nanomaterial that makes use of high performance technology while being effective, productive, and safe.

UNIVERSITY OF PUERTO RICO 2002

The University of Puerto Rico team had an obstacle that the other teams did not face—the Atlantic Ocean. To solve this problem they designed a house in four modules, loaded them on platforms, and hoisted them onto a cargo ship. Their structure was planned to manage most of its energy requirements with passive solar technologies. As was true of most of the other teams, the Puerto Rican students focused a lot of their attention on design that would "influence lifestyles and make an impact."

Green materials

- An operable metal shading device on the south facade reduces the energy spent on cooling by blocking heat gain.
- The roof has two main sections. One holds the solar powered energy system, and the other contains the solar hot water system.
- Materials used are all off-the-shelf to prove that "any house can go solar."

Green technologies

- The rectangular shape has narrow east and west dimensions, making up for lost space by elongating the north-south dimensions.
- A double swinging door leading off the bedroom to the west adds a nice touch.
- Fewer windows on the north side aid in passive solar energy techniques by blocking heat gain or loss.
- The students' desiccant and vapor compressor cooling system is a hybrid, using less energy than a larger one. The desiccant is included to remove humidity. Heat is produced when hot water passes through a heating coil in an innovative evacuated-tube system. A circular fin that fits around the perimeter of the evacuated tube acts as the absorbing material. A reflector underneath the tubes catches lost energy that escapes between the tubes, and the absorber material recollects it.

Figure 8.42   **University of Puerto Rico 2002—Solar Decathlon Project.**
(Image courtesy of U.S. Department of Energy (DOE)/National Renewable Energy Laboratory (NREL))

UNIVERSITY OF TEXAS AT AUSTIN 2002

The University of Austin's team entry was perhaps the most innovative of all the entries. As a team member put it, "Our design lets you take the core of your house to the beach with you for the weekend." The modular design starts with a large living space on the east which connects by a breezeway/deck area to an Airstream RV trailer docked in a bay in the middle of the house. Inside the Airstream are the kitchen and bath areas. One can walk straight through the Airstream into the bedroom module.

Green materials
- The steel frame is prefabricated.
- Structurally insulated panels (SIPs) are used as infill.

Green technologies
- Prefabricated, prewired modules allow efficient, rapid assembly. They also can be replaced or removed if desired.
- The photovoltaic system was designed by an engineering computer modeling program.
- The team bought two separately manufactured photovoltaic systems in order to be able to increase the number of panels that would fit on the roof. One system is only

Figure 8.43   University of Texas at Austin 2002—Solar Decathlon Project.

(Image courtesy of U.S. Department of Energy (DOE)/National Renewable Energy Laboratory (NREL))

used for charging their electric car (which is the contest-sponsored means of transportation donated to each team for the contest). The other system supplies energy for the house.

UNIVERSITY OF VIRGINIA 2002

The University of Virginia's entry was nicknamed the Trojan Goat. The amusing title was bestowed because like a goat it will eat anything (uses alternative energy). The exterior was very unusual in that it alternated narrow overhangs of wood with strips of copper cladding. The panels of wood stopped rain and shaded the copper panels. On the interior the smart wall, or nerve center of the house, was the most distinctive feature. The large wall was composed of light-emitting diodes that caused the wall to change color to correspond with the temperature in the house, something like a mood ring. The controls were automatic but could be overridden if desired, to select new colors or to have a constantly changing panorama of colors.

Green materials
- Reclaimed materials were used if possible.
- The deck was constructed of shipping pallets.
- A rain screen and louvered window coverings were made from reclaimed shipping pallets.
- The louvers used passive solar technology to block the sun when tilted down. When they were tilted up, they reflected light up to the ceiling of the living room. In winter, the louvers would open entirely to let in all the light.
- The wood used was birch and bamboo for their rapidly renewable qualities.
- A backyard garden featured vegetables planted in old tires.
- A green roof cut thermal gain and a roof device recycled rainwater to water the garden.
- The copper siding had formerly been a house roof.
- Engineered lumber, which uses a greater portion of the tree, was selected to construct the studs.
- The glass tile was all recycled.
- Reclaimed paving stones were selected.
- The recycled tires that acted as planters, when moved to their final location, would be reused in a retaining wall.

Green technologies
- A touch-screen computer controlled the house systems.
- A luminaire, a mirrored glass, tracked the sun and reflected light into a fiber optic cable, delivering natural light to the interior.
- Foam insulation gave the walls an R-50 and the roof an R-70 insulation value.
- Space heating had automatic controls and used a ground source heat pump with radiant floor heat.
- Space cooling was provided through a hydronic cooling system. The cooled air was transferred to the rooms via natural convection valances (vents above windows).

Figure 8.44   University of Virginia 2002—Solar Decathlon Project.
(Image courtesy of U.S. Department of Energy (DOE)/National Renewable Energy Laboratory (NREL))

VIRGINIA POLYTECHNIC INSTITUTE AND STATE UNIVERSITY 2002

The Virginia Polytechnic Institute team "took a significant stance to celebrate photovoltaic and not try to hide it. It's a benevolent umbrella," said the project team advisor. The interesting roof line of the house demonstrates what he means. The solar electric panels lift up at a 45° angle to shade the house as well as produce energy, giving the appearance of a row of fins marching across the roof. The multifunctionality of the panels matches that of other aspects of the house. For instance, the interesting Skywall panels on the facade, made of a translucent aerogel material, both insulate and allow light to pass through.

Green materials

- Skywall panels on the south wall are active solar collectors, used for producing hot water and powering the under-floor radiant heating.
- The interior north wall holds all the appliances, providing a thermal wall that retains heat in the winter and conducts it out in the summer.

Green technologies

- All appliances were placed on the north wall, which allows them to serve as a thermal buffer, retaining heat in the winter, but venting it out in the summer.
- A ground source heat pump aids in heating and cooling.
- Light is transferred throughout the house with a microprism light.
- Water is heated with absorber plates in vertical collectors.

Figure 8.45   Virginia Polytechnic Institute and State University 2002—Solar Decathlon Project. *(Image courtesy of U.S. Department of Energy (DOE)/National Renewable Energy Laboratory (NREL))*

Solar Decathlon 2005

Winner: University of Colorado at Boulder and Denver

CALIFORNIA POLYTECHNIC STATE UNIVERSITY 2005

California Polytechnic State University chose as their key words "simple" and "elegant." This was partly because of the great distance separating their university from the National Mall in Washington, D.C. (2394 miles). In fact, the students learned to call their entry "The One Truck Solution." While some entries had all the bells and whistles, this team wanted the inhabitants of their house to interact with it. Typical of California homes, this house has large double doors on the south opening out onto an oversized deck.

Green materials
- SIPs were used to construct the house.
- Exterior wood detailing was all of FSC lumber.
- Spray ICYNENE® insulation was used where there were gaps in the use of SIPs.
- Bamboo and linoleum flooring were used for their rapidly renewable qualities.
- The daylighting lit the house well and the electric lighting used accessible technologies and fixtures.

Figure 8.46   California Polytechnic State University 2005—Solar Decathlon
Project. *(Image courtesy of U.S. Department of Energy (DOE)/National Renewable Energy Laboratory (NREL))*

Green technologies

- Twenty eight solar photovoltaic panels produce 4.9 kW, easily enough to power the house.
- On sunny days, extra energy is stored in 24 batteries.
- The team also provided a system of hot water panels that collect heat from the sun and use it to heat water for the house. When this is insufficient, a thermal tank is used to heat water as needed.
- Overhangs on the south prevent heat gain during the summer but admit light in the winter.

CANADIAN SOLAR DECATHLON 2005 (CONCORDIA UNIVERSITY AND UNIVERSITÉ DE MONTRÉAL)

The Canadian Solar Decathlon team's vision was to design a house that looked like a traditional southern home with all the incorporated technology seamlessly hidden. Technology within the house included automation system monitors, which control the home's temperature and links to other controls throughout the house. The house also used traditional southern cooling methods like the wrap-around porch which provides much-needed shade. The roof looks clean and uncluttered with the photovoltaic panels blending in unnoticeably. One interesting innovation is the window with solar electric cells and an autumn leaf design. Though the house will need to endure a harsh climate, the large southern windows open up the vista and are triple glazed for winter weather.

Figure 8.47   **Canadian Solar Decathlon 2005—Solar Decathlon Project.**
(Image courtesy of U.S. Department of Energy (DOE)/National Renewable Energy Laboratory (NREL))

Green materials

■ Phase change materials are substances with a high heat fusion which melts and solidifies at a certain temperature and is capable of storing and releasing large amounts of energy. Phase change materials and a wall of water placed adjacent to the window glass serve as thermal storage and continue the house's invisible technology design approach.

■ Solar panels on the roof are efficient but unobtrusive.

■ One window has solar electric cells which are similar to photovoltaic panels and are capable of changing color and making designs.

Green technologies

■ Warm air from the house solar panels, which would normally dissipate, will find a use in the clothes dryer, one of the biggest electrical loads in a house.

■ Control-operated window blinds open or close window blinds according to the amount of heat the system deems necessary to help maintain a predetermined temperature.

CORNELL UNIVERSITY 2005

The Cornell University team kept their eye on the eventual purchaser of the home. They decided on a modular design so the house could grow along with the owner and his family. Likewise, the student design team wanted to create a home that could be mass produced. They even designed alternate versions—some that could stand up to colder climates. Their ultimate goal was to produce a great design that can be sold for $50,000 to $100,000. The house is long, running on an east to west axis.

Green materials

- Crystalline-silicon photovoltaic panels provide power for the heat pump and the home's electricity.
- Two energy-efficient lighting strategies are natural daylighting and an electric fixture that reflects light onto the ceiling. This strategy amplifies the effect of the electric light but uses a smaller amount of electricity.
- Cork flooring was used.
- The light canopy supports for photovoltaic, evacuated tubes for solar water heating, and vegetated screens that provide shade all help to lower the amount of electricity being consumed.
- Bamboo was chosen for the cabinets in order to reduce the VOCs.

Green technologies

- East-west axis positioning allows the sun to take the major role in lighting the building.
- On the roof water needs are met by evacuated tubes. Hot water is used for radiant flooring system.
- Crystalline-silicon photovoltaic panels power a heat pump and the electricity.
- The students custom-designed an energy recovery ventilator that has as the major component a rotary wheel composed of silica gel, which is an excellent absorber of humidity. In summer the silica gel takes moisture out of the fresh air intake before it reaches the house. It transfers the humidity to the air being exhausted along with the extracted heat.
- Water systems, collection of heat for hot water, water filtration, and reuse of water are consolidated within a central strip.
- The living space is separated from the sleeping area by a core of water systems and mechanical systems.
- Gray water and rainwater is collected and filtered through an architectural water element in the landscaping. This "water wall" is sloped so that gravity allows the water to travel along it. By the time the water reaches the end of the wall it is safe for use in the landscaping.
- Appliance and fixtures are set up together to minimize extra piping.
- Separation of spaces is generally avoided to give flexibility to the owner.

CROWDER COLLEGE 2005

Crowder College took their inspiration from architect greats like Frank Lloyd Wright. The design is reminiscent of the beautiful but simple Arts and Crafts Movement that is still with us today in the term sustainable design. The simple and pleasing house is primarily wood, both in the exterior and in the built-ins that the style is known for. This old style received a modern touch with the "pods" or modules that compose the

Figure 8.48　**Cornell University 2005—Solar Decathlon Project.**
(Image courtesy of U.S. Department of Energy (DOE)/National Renewable Energy Laboratory (NREL))

students' house and that detach and fold down when being disassembled. The bunga-low style is enhanced by the small courtyard on the north. The large photovoltaic solar thermal system on the south is exactly positioned to "catch the rays."

Green materials
- Wood from the Pioneer Forest, the only forest certified as responsible and renewable forest management, is being used in the flooring and cabinetry.

Green technologies
- A radiant floor heating method allows the building to have its heating needs met by running tubes containing hot water through the floor. Interestingly enough, the water that is used comes from a hot tub on the back deck.
- Their passive solar heating system uses the sun to warm the water for the laundry, kitchen, and bathroom. This process cuts down on a lot of energy consumption by using a passive system instead of a typical water heater.
- Practicing reuse, the students are using the 2002 Decathlon trailer, which holds a battery system run by photovoltaic power, for the contest.

Figure 8.49   Crowder College 2005—Solar Decathlon Project. *(Image courtesy of U.S. Department of Energy (DOE)/National Renewable Energy Laboratory (NREL))*

FLORIDA INTERNATIONAL UNIVERSITY 2005

Florida International University designed a house that fits perfectly into the hot Florida climate. The exterior appears to be all glass—from the huge double doors to the extremely large windows. In fact the Florida house is about one-third hurricane-proof glass. The house is named Engawa, a Japanese word that deserves a little explanation. Engawa describes a space that is both inside and outside. If one looks at the front of the house the large glass expanse seems to be in the shape of the letter U. FIU's decathlon house is hurricane compliant, utilizes sustainable materials and technologies, including solar water-heating system, photovoltaic-integrated windows, ductless air conditioning, bamboo flooring, and bioplastic lumber decking.

Green materials
- Bamboo flooring.
- LED lighting.
- Bioplastic lumber decking.
- Since the building is over 30 percent glass the challenge is to keep heat out. This is achieved with operable windows and blinds.

Green technologies
- High-SEER ductless air-conditioning system.
- TCT Solar provided the PT-50, a 50-gal ProgressivTube solar water-heating system.

Figure 8.50   Florida International University 2005—Solar Decathlon Project.
(Image courtesy of U.S. Department of Energy (DOE)/National Renewable Energy Laboratory (NREL))

■ An interesting variation is that some of the windows have photovoltaics integrated into them. The windows of this type can also be used as a projection surface if the owner chooses to display visual imagery on them.

NEW YORK INSTITUTE OF TECHNOLOGY 2005

The New York Institute of Technology team decided to call their entry Green Machine-Blue Space, with a theme of regeneration and reuse. On the left side of the front facade is a recycled green shipping container (the Green Machine part of the house). On the right is the rest of the house—Blue Space, the living area, which was constructed of wheat straw panels. In the Green Machine are the "guts" of the house, the power users and heat-producers. There one will find the kitchen and bathroom, topped by a green roof that produces food for the table. Blue Space, on the other hand, is the living space, open and filled with light. Above, a sleeping loft connects to the roof garden on Green Machine by means of a bridge. The ground link between the two spaces is a breeze-way. One can walk straight through from the front yard to the backyard through this feature. The team thought long and hard about treading lightly on the earth. They were proud of the fact that their entry lives in complete harmony with its environment.

Green materials
■ Soybean insulation is an insulation that is water-blown and soy-based, which seals a structure's thermal envelope, making it more energy efficient, healthier, more comfortable, and durable than traditionally insulated homes.

Figure 8.51   New York Institute of Technology 2005—Solar
Decathlon Project. *(Image courtesy of U.S. Department of Energy (DOE)/National
Renewable Energy Laboratory (NREL))*

■ Agriboards are structural insulated panels (SIPs) made from wheat straw. The
wheat straw building panels are recyclable solid panels made of all natural fibrous
raw materials. Their use speeds up the construction process. The panels can be used
for load and nonload bearing uses.

■ Each piece of furniture is multifunctional with some unique features; for instance,
the sofa will warm or cool its occupants.

■ LED-powered task lighting was chosen.

■ Double-paned, argon-filled windows offer more insulation to the building.

Green technologies

■ The 55 panels of photovoltaic cells are mounted on the roof of Blue Space, where
the cells convert sunlight into electricity by separating water into hydrogen and
oxygen through electrolysis.

■ The energy is stored in a hydrogen energy storage system. The hydrogen is then fed
into a fuel cell where the energy is produced by recombining hydrogen with oxygen. One student put it this way: "We're taking energy from the sun to produce
hydrogen from water, and turning that hydrogen back into water to produce electricity. It's a clean, renewable, and elegant cycle."

■ A solar-powered evacuated-tube system was used to heat domestic hot water.

PITTSBURGH SYNERGY 2005 (CARNEGIE MELLON, UNIVERSITY OF PITTSBURGH, ART INSTITUTE)

The Pittsburgh Synergy team members have a house that seems to be all about the sun. The
north and south walls, which tilt 12° to the south, give the house an "all angles" look. The

south wall is completely made up of windows; the bottom of the highest row of windows can be tilted out. It's a great way to let the outdoors in, and in the winter the sun beaming into the living and office space can lift anyone's spirits. The glare and heat is controlled with manually operated shades. The floor of the living/office (or great room) is concrete, which stores heat from the sun and contains radiant heat pipes to make it cozy during the winter; conversely the thermal mass of the concrete feels cool in the summer. There is truly a sense of being outside when one is sitting in the interior of the house, and the deck with comfortable seating just outside the south facade seems to bring the living area outside as well.

Green materials

- The tilted north wall is composed of sheets of polycarbonate embedded with glass beads. Polycarbonate is plastic, which is extremely strong as well as being translucent.
- The translucent north wall lets a diffused light into the bedroom while keeping it private.
- SIPs with wood cladding make up the west and east walls.
- LED lighting is seen throughout.

Green technologies

- The mechanical and private spaces are located on the north side of the house, while the more public spaces are on the south.
- Most of the available solar input is used to heat water. The hot water is not only used for domestic purposes, but also for radiant heating and for cooling.
- Absorption chilling is the method used to cool the house; it is a refrigeration technology using heat rather than mechanical compression.

Figure 8.52   Pittsburgh Synergy 2005—Solar Decathlon Project.
(Image courtesy of U.S. Department of Energy (DOE)/National Renewable Energy Laboratory (NREL))

RHODE ISLAND SCHOOL OF DESIGN 2005

The Rhode Island School of Design's entry is perhaps the home that most clearly moves away from the traditional and focuses on high tech design. It was intended to be at home in an urban setting, perhaps in a line of row houses. Descriptive terms such as "loft-like" and "free-space" have been used. A louvered skin is one of the unusual features of the home. It is, of course, adaptable—able to keep heat in when the weather is cold and reflect heat when the temperature climbs. Even more interesting is the ability of the skin to provide color variations (rather like a chameleon) through a graphic use of hidden colors as the earth spins and the skin tracks the sun's movement.

Green materials
- Twenty four Sanyo photovoltaic panels were chosen, producing 190 W for each panel.
- A large wall of windows in the living room can slide open so that the interior merges with the balcony deck.

Green technologies
- The north-south orientation is adaptable to the city, as it positions the home to receive daylight even when in a row of townhouses.

Figure 8.53   Rhode Island School of Design 2005—Solar Decathlon Project.
(Image courtesy of U.S. Department of Energy (DOE)/National Renewable Energy Laboratory (NREL))

- The mechanical core of the home is stacked and centrally located, which cuts down on the need for extra pipes and ducts throughout.
- Batteries hold a three-day's supply of power.
- The roof garden is a more traditional touch, and very pleasing in an urban setting where there may be no garden or lawn. Vegetables and herbs will be grown in planter boxes.
- The construction of the house is modular.

UNIVERSITY OF PUERTO RICO 2005

The home of the Universidad de Puerto Rico's team is best described as hospitable. They truly capitalized on the "mi casa es su casa" proverb. The beautiful deck with its central rail seeming to invite visitors to enter typifies this motif. And if one needed to make it even more open and expansive, consider the living area which opens up to an outdoor deck when weather permits. All this is representative of the Puerto Rican culture. (We wanted) "to represent who we are," one team member remarked.

Green materials
- Luminaries light the way to this inviting house.
- Horizontal design elements with a shallow pitch to the roof add to the message of hospitality.

Green technologies
- The clothes dryer was modified to dry items with hot water heat rather than electricity.
- The large east deck corresponds to the friendliness of the southern deck area.
- Mechanical functions take place on the west side of the house in a separate module.

UNIVERSIDAD POLITÉCNICA DE MADRID 2005

The Universidad Politecnica de Madrid was the only European entry in the Solar Decathlon in 2005. Like some of the other teams, the Madrid students did not want to have a house that was mainly about gadgets. Attractiveness and comfort were the primary design objectives. Students describe it as "a flexible Mediterranean-style home with a great interior-exterior connection." Versatility is the feature that stands out the most. It was called the "magic box" because of its versatile spaces. A set of moveable walls allows it to be divided into three spaces, five spaces, or to be completely open. The huge glass facade on the south allows the living room to become an interior courtyard or patio.

Green materials
- Fluorescent lamps, as more efficient than halogen bulbs, are used on the interior of the louver-door countertop.
- Ceramic flooring holds heat or coolness better than carpet.
- Bioclean glass is used for greenhouse windows.
- Low-E glass is used in all panes of fixed glazing, and the kitchen window is made of glass with thermal properties. The living room has windows with solar-electric cells.

Figure 8.54 Universidad Politécnica de Madrid 2005—Solar Decathlon Project.
(Image courtesy of U.S. Department of Energy (DOE)/National Renewable Energy Laboratory (NREL))

Green technologies
- The external side of the photovoltaic wall is a double glazing, half inch thick where the photovoltaic cells are contained.
- The main source of daylighting is the windowed south wall.
- Electrochromic windows darken or lighten as necessary.
- Phase-changing gels in the foundation are used for thermal control.
- LED lighting is used on the outside of the house.
- There are two solar greenhouses.
- Evacuated tube solar collectors on the roof are oriented for optimal sun collection.
- Three windows can be opened to provide natural ventilation.
- A solar water heating system heats the water for the house.

UNIVERSITY OF COLORADO AT DENVER AND BOULDER 2005

The BioS(h)ip is the name of the entry from the University of Colorado. That is mainly because the innovative SIPs used to construct the walls of the house were developed right there at the University and were named BIO-SIPs. The new product is a composite of Sonoboard, produced by Sonoco Company and composed of recycled cellulose materials, and BioBase 501, a foam insulation made of soybean oil. The foam insulation is placed between two Sonoboard panels, like an "ice cream sandwich." The merging of the two products makes a lightweight SIP and has been patented.

The materials which make up the BioS(h)ip have strangely edible components. Soy, corn, wheat, canola, coconut—these are just some of the natural products that have been factory-pressed and molded into panels, building products, bricks, etc. These biobased products have no petroleum or very little.

The design of this entry is different from the 2002 home. This house is more modern and innovative and strives to be a "modular home for the future."

Green materials

- Using natural materials was one of the team's five major design goals, along with innovation, energy efficiency, modularity, and accessibility.
- Plastic dishes made of cornstarch are used in the kitchen.
- Bamboo was chosen for kitchen shelves.
- Recycled paper wall panels were used.

Green technologies

- Large windows on the north provide plenty of diffused rather than glaring light but do not have the challenge of combating heat gain and loss.
- High-performance window glazings are used, and clerestory windows brighten the kitchen.
- An integrated radiant-solar thermal system is used for space and water heating.
- The rooftop photovoltaic system is made of 34,200 W panels. Photovoltaic cells are also combined with the overhangs that shade the southern facade.
- Clerestories are made of translucent aerogel-filled panels.
- Transportation of the building was accomplished by naturally derived fuels.

Figure 8.55   **University of Colorado, Boulder and Denver 2005—Solar Decathlon Project.** *(Image courtesy of U.S. Department of Energy (DOE)/National Renewable Energy Laboratory (NREL))*

UNIVERSITY OF MARYLAND 2005

The University of Maryland wanted a design that did not look anchored to the ground, but that appeared to be "floating." They decided to support the house with pedestals resting on blocks of stone. This allows the earth under the house to remain more undisturbed. It also had the added advantage of providing storage space for batteries and water tanks. The roof is constructed with curved joists which give a beautiful shape to the house and, in terms of energy, facilitate the sun's rays reaching the photovoltaic panels more readily as the year advances. Choosing FSC lumber and bamboo flooring emphasized the team's commitment to rich, comfortable, but at the same time healthy, building principles.

Green materials
- All the lumber used was FSC certified.
- Concrete floors are heated by a radiant heating system, although the thermal mass of the concrete picks up heat from the rays of the sun entering through the large windows.
- Low-VOC paints were used as a healthier alternative to traditional paint.
- Windows are triple-paned and operable for natural ventilation.

Green technologies
- Fifty one charcoal-gray photovoltaic panels, along with curved roof joists, make an efficient but aesthetically pleasing roof.

Figure 8.56   University of Maryland 2005—Solar Decathlon Project.
(Image courtesy of U.S. Department of Energy (DOE)/National Renewable Energy Laboratory (NREL))

■ PVC-free plumbing, a recent invention and technological advance which is designed to prevent the eventual leaching of harmful chemicals into the environment, was used.

UNIVERSITY OF MASSACHUSETTS AT DARTMOUTH 2005

The University of Massachusetts at Dartmouth began with the goal to produce a home that could function with natural energy for a long life span. Before they had pounded the first nail, they had planned for the final location of their project—as a Habitat for Humanity home in northeast Washington, D.C. The team of 16 had previous experience working for Habitat for Humanity in Tennessee a year prior to the Solar Decathlon 2005 contest. Because of this, they wanted the house to be as "normal-looking and comfortable" as possible so that it would fit right in to the neighborhood where it would eventually be transported.

A computerized control system is used in the home to regulate phase-change materials. These materials, kept in storage under the front windows, are made from blocks of cloth. Al Rossetto, an ecodesigner from Vermont, consulted with the team to guide them in following the five essential principles of building an energy-efficient home: (1) the foundation must be insulated, stable, and ensure adequate drainage, (2) SIPs to frame and insulate the home, (3) multipaned windows with sealed frames, (4) radiant heat to heat the house, and (5) a heat recovery ventilator (HRV) to balance ventilation in the home.

Green materials
■ Energy star appliances were donated by Whirlpool.
■ Recyclable materials for economical construction were used whenever possible. For example, the SIPs (structurally insulated panels) use recycled material to insulate the home.
■ Two photovoltaic systems collecting power from the sun are housed on the roof.

Green technologies
■ The house uses forced-air heat exchange: The phase-change material, mentioned above, which passes intake air, and secondly, radiant-floor heating. The piping in the floor receives hot water by one of the evacuated-tube solar thermal collectors housed on the roof of the house. The other energy collector is used to provide hot water.
■ Lighting, also, is achieved by two methods: Compact fluorescents and natural daylighting, which is facilitated by light tubes/chimneys bringing light down from the roof.

UNIVERSITY OF MICHIGAN 2005

MiSo (Michigan Solar house project) is the name of the University of Michigan team's entry. As is true of some of the other teams, the Michigan students wanted less tech and more familiarity to characterize their home. The Michigan team did not want viewers to think, "That's a real neat idea, but I wouldn't want to live there." As with the Massachusetts team, this group of students was thinking long life span. They stated that their building should last for 100 years.

The shape of the structure and the envelope of the building are immediately noticeable. Made of aluminum as the cladding, the external material actually helps support some of the weight of the house. Though not particularly energy efficient, aluminum is a durable material and retains its value when recycled. The shape is rounded from the roof to the ground, with photovoltaic panels following the curve of the design. An interesting energy idea is the glass spaces at the base of the southern facade. These are heated by the sun's rays, and this heated air rises along the curved facade of the house by means of a solar chimney. In the winter, this heat enters the house through louvers, warming the interior. In the summer the heat is released into the outside air. Whichever process is taking place, this layer of air within the glass spaces is a good insulator.

Green materials
- Plenty of arching windows follow the roof line and take advantage of the passive solar approach.
- A solar chimney brings light and heat down from the roof. The curve of the house facilitates air rising up and around the design of the house.
- Sunflower board and ash are used in the interior; they are recycled and easily recyclable.

Figure 8.57   **University of Michigan 2005—Solar Decathlon Project.**
(Image courtesy of U.S. Department of Energy (DOE)/National Renewable Energy Laboratory (NREL))

Green technologies

- The design utilizes solar energy by placing solar panels all around the curve of the building.
- The arched, curved wall combines insulation, ventilation, light reflection, and help bear the weight of the structure.
- Solar panels control the radiant heated floor.
- Excess energy is stored by batteries located in the floor.
- The house is lightweight and easily transported.

UNIVERSITY OF MISSOURI-ROLLA AND ROLLA TECHNICAL INSTITUTE 2005

The "Prairie Home" entry by the University of Missouri-Rolla and Rolla Technical Institute expands the use of solar energy to make more visually acceptable. Accordingly, their house would be right at home in the traditional Midwest. Since Midwesterners are known to eschew anything different, that is an important feature of the building. The northern portion of the roof has the same pitch as the southern portion. This avoids that photovoltaic look.

One aspect of the house that is visually appealing, but probably not apparent to the untutored eye, is that everything in the house is designed around the Fibonacci Sequence (the Golden Ratio). This mathematical sequence can be found in patterns in nature in anything from leaves on a tree to the way seeds are distributed in a raspberry. The first plan for the house was a shell-shaped facade as illustrating this sequence, but it was later decided it would not be traditional enough. A work of art created by a student and hung on a wall is just one illustration of how this sequence is carried out in the house.

Green materials

- The standing-seam copper roofing is highly resistant to corrosion; after it has been left to oxidize, a patina (turquoise) finish results. The material works well in places with frequent earthquakes because it is very lightweight, and in places with a lot of snow, it allows for snow removal very easily. It is also resistant to mold and fire resistant.
- SIPs create a durable and thermal exterior protection.
- Wood detailing and paneling is found throughout the house, even in the bathroom.

Green technologies

- The curved island in the kitchen area follows the Fibonacci principle.
- With twenty-eight windows and two doors, daylighting is maximized, reducing the need for electrical light.
- The living area is in the central portion of the house in a taller module with an angled roof. Inside, this feature provides a beautiful wood-paneled ceiling, and outside, solar energy collection is optimized.
- A solar thermal and electrical photovoltaic panel system increases efficiency, effectiveness, and productivity.

Figure 8.58   **University of Missouri-Rolla and Rolla Technical Institute 2005—Solar Decathlon Project.** *(Image courtesy of U.S. Department of Energy (DOE)/National Renewable Energy Laboratory (NREL))*

UNIVERSITY OF TEXAS AT AUSTIN 2005

The University of Texas at Austin team members felt their home had two purposes to fulfill. One was to build a home that not only sustained itself with its own power but actually produced energy. And secondly, they saw it as their mission to preach the importance of sustainability to the general public.

The name of their entry is called SolarD SNAP House (Super Nifty Action Package House). Like Lego blocks the modules are prefabricated and "snap" together for simple assembly and transportation. Team members worked the module design idea into every detail. They even came up with an innovative strategy for assembling the roof.

Green materials

■ The wood used in the project is reclaimed redwood and mesquite wood, the latter being a common Texas wood. The reclaimed redwood is used for the screening of the south window of the house, allowing a limited amount of sunlight to be directly transferred into the house. The mesquite is a recycled material that makes up most of the interior.

■ Another eco-friendly aspect of the house is the use of expanded polystyrene foam core for the insulation of the wall panels.

■ The green roof is planted with grasses native to Texas, which grow taller than lawn grass. It adds insulation from heat gain.

■ Forty two photovoltaic panels are installed on the south side of the roof. The panels were awarded them as a prize from BP Solar, since the company felt that of all the contestants, this team had done the best job of integrating solar panels into their home. The solar collectors even form a design pattern on the house.

■ The exterior of the house is made of recycled zinc cladding, a good insulator and also a durable material.

■ A Polytronix skylight controls the light entering the bathroom.

■ Warmboard radiant heat subfloor was chosen.

Green technologies

■ The computer in the house runs a program that monitors and controls the environmental control systems in the house 24 h a day. The computer controls the output of the energy throughout by turning off items, such as itself, when they are not in use.

■ A translucent ecoresin is used as a finish for all of the panels.

VIRGINIA POLYTECHNIC INSTITUTE AND STATE UNIVERSITY 2005

Perhaps the most interesting feature of the Virginia Polytechnic Institute and State University home is that it arrived at the National Mall in one piece. It actually is a mobile home as well as a stationary home. According to the students' faculty advisor, the team wanted "to be able to transport our home to Washington intact so we could spend the five days in D.C. fine-tuning it." The transportation process is quite unique. The home itself is a lowboy trailer attached to a tractor with a detachable connection. The tires on the home are also detachable after the project arrives at its destination. The house is supported with trusses as it travels, and these trusses can fold down to become supports for the deck of the house.

A second outstanding feature of the home is the translucent walls with LED lighting near the foundation that makes the house glow at night. The roof, too, curves upward somewhat like a shallow letter V reaching for the sky. The underside of the roof reflects sunlight into the house.

Green materials

■ Three sides of the house were constructed with thin layers of translucent polycarbonate material, filled with aerogel insulation.

■ SIPs form the north wall, which is where the appliances and energy systems are located.

■ Bamboo flooring is used on the interior.

■ Linoleum forms the flooring for the mechanical and appliance room.

Green technologies

■ Hot water panels heat water. Two Heliodyne liquid flat plate collectors are used to provide the house with hot water. A heat transfer fluid is circulated through the collectors where it absorbs energy from the sun. This heat is then transferred to the water in the storage tank.

Figure 8.59   Virginia Polytechnic Institute and State University 2005—Solar Decathlon Project. *(Image courtesy of U.S. Department of Energy (DOE)/National Renewable Energy Laboratory (NREL))*

- The pump shuts off at night and the fluid from the collectors drains back to the drainback tank. This prevents freezing of the heat transfer fluid in the collectors.
- A thermal storage tank stores the hot water produced by the solar thermal system during the day for use at any time. The storage tank also has an electrical heating element that can be used to heat the water when energy from the sun is not sufficient.
- The faculty advisor Bob Schubert describes the structure as the polycarbonate walls and the roof form a tunable enclosure system that can be adjusted daily for a particular location or used to adapt to many different climate zones.
- Motorized shades provide privacy at night; during the day they help maintain a stable temperature.
- A Trombe wall located on the south side provides thermal heat storage. The wall absorbs heat and retains heat to maintain the house's temperature.

WASHINGTON STATE UNIVERSITY 2005

The Washington State University team's goal was to create a house that featured the "green" that seems iconic of the northwest. This image was exemplified by the two new bills passed by the state that aggressively promote solar power. The core of the house is a used shipping container, one of the new materials of affordable housing that has become so popular lately. This provides easy access to all the mechanical systems.

A new SIP, designed by a Spokane architect, is featured in the home. Normally a traditional panel consists of a foam core with a facing glued on both sides. The most common facing materials are plywood and oriented strand board. In this innovative SIP, the core is a corrugated steel frame, and this frame is encased in polystyrene material. In this new type of SIP, plumbing and/or electrical wiring can be run through the panel without weakening it.

Green materials

- The refrigerator was purchased from a marine supply store. It is so efficient that it need run only 1/20 of a day. This option allowed the team to modify the refrigerator compressor to also run the air conditioning for a major share of the day.
- Concrete floors create thermal mass.
- A corrugated steel frame makes a strong durable structure.
- Skin, decking, and siding are all made up of wood-plastic composites.
- Industrial-style metal sheeting was used as shading devices.
- Eucalyptus flooring is used for interior floors.

Figure 8.60   **Washington State University 2005—Solar Decathlon Project.**
(Image courtesy of U.S. Department of Energy (DOE)/National Renewable Energy Laboratory (NREL))

Green technologies

- Radiant heating is used under the concrete floors. They also hold heat that passes through the large windows.
- The beams in the house are made from parallel strand lumber, which are long strands of wood laid out parallel to each other and then glued together with an adhesive. They can bend without breaking. Ecologically this type of wood saves old-growth forests which would otherwise be cut down to produce these long beams.

Solar Decathlon 2007

Winner: Technische Universität Darmstadt, Germany

CARNEGIE MELLON 2007

Tripod is the name the Carnegie Mellon team gave their solar home. The name is derived from the three "legs" that comprise the three main living spaces in the structure. The core structure, a long metallic rectangle, houses all the mechanical, electrical, and water systems, as well as a laundry area and a bath on the far eastern end. The core is a hallway which is not as finished as the three legs, and gives one the impression of being in a machine. The kitchen is the leg jutting out to the north of the core, and the bedroom area and living area are the two remaining legs, respectively, thrusting out to the south. The whole concept is a "plug and play" philosophy of architecture, which retains only the core as permanent, while the legs can be detached or added as the number of occupants in the home increases or decreases. But the house is not just about the interior. Two added features bring the outdoors into the design—the greenscape on the north and the southern courtyard.

The student designers planned the building based on three main focus areas: plug and play (indicating the modules which can be detached or added); sustainability as a global concept; and designing for an exhibit (since the house eventually will be placed in the Powdermill Nature Reserve as an educational resource).

Green materials

- Solar thermal tubes double as a skylight to let natural daylight into a dark space.
- Daylighting, instead of electric lighting, is used as much as possible.
- High-efficiency insulation retains the desired temperature in the house.
- White oak from Pennsylvania was used on the exterior.

Green technologies

- The house is broken up into zones, which avoids one central setting for the entire house, saving energy that would normally be used in unoccupied spaces. Each pod has its own individually controlled air conditioning and radiant floor heating system.
- Solar thermal collectors above the bathroom absorb heat from the sun, which is transferred into a water supply to provide domestic hot water.

Figure 8.61   Carnegie Mellon 2007—Solar Decathlon Project.
(Image courtesy of U.S. Department of Energy (DOE)/National Renewable Energy Laboratory (NREL))

- A large water tank is filled from the downspouts from the roof, and the collected rainwater is used in tandem with a drip-system to water the plants on the site.
- The home utilizes a passive heating system with windows on the southern face. Cool air is drawn in through lower windows, while hot air rises and escapes through the clerestory.
- A solar electric system of 6.88 kW generates electricity that is stored in a large battery bank.
- The house is also cooled by a "greenscape," plants that grow up the walls and even onto the roof.

CORNELL UNIVERSITY 2007

The entry from Cornell University is remarkably streamlined in appearance. The dominant feature is the separate steel solar canopy, which gives the advantage of being able to modify, interchange, or update the components without interfering with or damaging the building envelope. Furthermore in the space between the canopy and the roof of the building, landscaping and/or shading devices can be placed.

In the house itself, the living area is on the far east; a large sun room occupies the central portion, followed by the bedroom. On the north are the kitchen, bathroom, and mechanical closet.

Figure 8.62 Cornell University 2007—Solar Decathlon Project.
(Image courtesy of U.S. Department of Energy (DOE)/National Renewable Energy Laboratory (NREL))

Green materials
■ Raised access flooring was designed; this allows adjustments to be made, for instance, to wiring and ventilation systems.

Green technologies
■ Steel trusses frame the solar canopy and serve as photovoltaic support.
■ Evacuated tubes for solar water heating are found in the solar canopy.
■ Under the solar canopy are the vegetated screens that shade the house in the summer.

GEORGIA INSTITUTE OF TECHNOLOGY 2007

The Georgia Tech team must have been thinking "Let's bring in the sun," for that is what their design accomplishes. One team member remarked, "We've placed a great emphasis on light and bringing light into the house in unique ways." One outstanding example of how the team has accomplished this is the translucent walls and roof.

Green materials
■ Biobased insulation is insulation derived from soy beans.
■ A row of clear windows above the walls of the house, the clerestory windows, continue the theme of bringing in the daylight. They also provide views of treetops, which seem to bring the outside into the living areas.

- The translucent walls are constructed with panels made of two sheets of polycarbonate with a layer of aerogel between the sheets.
- Aerogel is the lightest solid known, and not only insulates well but also has the advantage of allowing filtered light to pass through it.

Green technologies

- The roof of the building, which is made of translucent film, also transmits light. The two layers that make up the roof are: an aerogel layer for insulation and a top layer designed to drain the roof and shed water.
- The rain screen on the exterior of the house protects the envelope from moisture that leads to decay.
- The photovoltaic panels on the roof are adjustable and shade the house from direct sunlight.
- Horizontal louvers enclose the south and east walls to shade the translucent walls from direct sunlight.
- The floor is of light frame wood construction with a bio-based insulation sprayed between to make the floor air tight.
- An energy recovery ventilator conditions incoming air while the exhaust air rotates a wheel, thus recovering energy through a pollution-free system.
- Solar lights are used inside because they are self sufficient, with a battery for use during the night.
- Photosensors are used to sense movement and activate lights.

Figure 8.63   Georgia Tech 2007—Solar Decathlon Project. *(Image courtesy of U.S. Department of Energy (DOE)/National Renewable Energy Laboratory (NREL))*

KANSAS PROJECT SOLAR 2007 (KANSAS UNIVERSITY AND KANSAS STATE UNIVERSITY)

The Kansas University and Kansas State University collaborative team stated as their mission: "to capture the essence of Kansas—its landscape, history, economy, and resources." They also planned to embrace the challenge of transporting their home with an entire mobile concept. The fully assembled dwelling is able to fit on one truck.

The narrow profile of the building shows off the energy saving features. The solar panels cover part of the south facade and are tilted at 64° to take advantage of winter sun. Roof panels can track the sun's path to increase their ability to capture more energy.

Green materials

- SIPs of two outer layers of plywood and an inner insulating core of extruded polystyrene were used.
- Reclaimed barn-wood was used in both exterior screening and interior cabinetry.
- Glass counters and floors were made from post-consumer recycled glass products.
- Insulated glazing is a high-performance material since it provides better insulation than equivalent alternatives, letting in light without normal heat loss.
- Ash and cypress were used in the construction of this project. These woods were chosen because they could be harvested locally, thus reducing the embedded energy

Figure 8.64   **Kansas Project 2007—Solar Decathlon Project.** *(Image courtesy of U.S. Department of Energy (DOE)/National Renewable Energy Laboratory (NREL))*

of the material. In addition, students bought wood only from FSC certified forests, to promote sustainability.

■ The students chose Bosch appliances because of their sleek contemporary look and high efficiency. For example, an induction cooktop only heats the kettle and food; the stovetop does not get hot to the touch.

■ NanoGel is a part of the skylights because of its excellent insulation.

■ Biodegradable trays were used when constructing the greenroof rather than recycled plastic.

Green technologies

■ The photovoltaic system is made up of BP Saturn 7180 panels arranged along a sloped southern facing wall; the system includes sealed batteries.

■ The living space can be expanded onto the deck area.

■ Batteries hold three days of power.

LAWRENCE TECHNOLOGICAL UNIVERSITY 2007

The Lawrence Technological University entry is named Aloeterra, reflecting its healing effect on the earth.

The design of the building shows a recessed doorway/court which opens onto a hallway, at either end of which is a bathroom and a bedroom. The kitchen, dining, and

Figure 8.65   Lawrence Technological University 2007—Solar Decathlon Project.
(Image courtesy of U.S. Department of Energy (DOE)/National Renewable Energy Laboratory (NREL))

living spaces are on the west, while the bedroom is on the east. The technical and machinery section is located on the north wall.

Green materials
- Sustainable local materials were chosen.
- Decking was built from a composite of rice hulls and polymer.
- Both LED and fluorescent bulbs were used.

Green technologies
- A central solar chimney aids in directing heat and light into the house.
- The roof is installed extensively with 36 solar electric panels.
- On the west side, is a "fence" extending from the front of the house and made up of solar thermal collectors.

MASSACHUSETTS INSTITUTE OF TECHNOLOGY 2007

Massachusetts Institute of Technology (MIT) entered the Solar Decathlon for the first time in 2007, but they were by no means novices to solar-powered home design and construction. In fact, the team calls their home Solar 7, as MIT has constructed six previous solar homes from as far back as the 1930s. Furthermore, MIT carefully analyzed the past entries in the Solar Decathlon for their strengths and weaknesses before beginning their own designs.

Several interesting features stand out in this building. A touchscreen computer coffee table is linked with software controls for utility systems in the house. And a Trombe wall made up of translucent tiles aids in transferring heat to the interior. A very visually welcoming feature is the generous deck with ramps for accessibility.

Green materials
- A south-facing thermal exchange wall also supports overhanging photovoltaic panels.
- An interactive information wall is an interesting feature of the house.
- The living area opens out onto outdoor deck areas.
- On the south wall is a portion containing "warm light" blocks of one-foot-thick square tiles, which transmit heat in the evenings and through the night.
- Energy-efficient windows have three panes of glass and an insulator of krypton gas.
- All of the material on the structural panels can be recyclable.
- Wood is used on the external envelope of the building.

Green technologies
- Photovoltaic panels generate about 9 kW per hour.
- Twenty four batteries store energy and can power the house for nearly 48 h.
- A translucent trombe wall allows the building to passively capture solar energy throughout the day.
- Sixty solar thermal evacuated cylinders support domestic hot water and the radiant floor heating.

Figure 8.66 **MIT 2007—Solar Decathlon Project.** *(Image courtesy of U.S. Department of Energy (DOE)/National Renewable Energy Laboratory (NREL))*

■ Along the north wall are the bedroom, office, and a large deck, which extends along the east and south sides of the house. The front of the house holds the bath, kitchen, dining, and living areas.
■ A molded subfloor embedded with plastic tubing holds the radiant heating system.
■ A waste mitigation system makes it possible to dispose of some waste before going to a landfill.

NEW YORK INSTITUTE OF TECHNOLOGY 2007

The New York Institute of Technology's "Open House" is aptly named as it is entirely open to the light and breezes on the south side. Everything in the house helps to maximize natural light and the outdoors: the white ceiling, clerestory windows all around the house, and a contained pond atop the roof being just two examples.

Green materials
■ Caesarstone, which is crushed quartz with a mixture of polymer resins compacted into slabs, is used for kitchen countertops.
■ A building panel called NanaWalls frames the south side.

Green technologies
■ Automated building controls include a special high-tech touch—a fingerprint scanner.
■ Open space in the house is about 480 ft^2, while enclosed core space is 270 ft^2.

Figure 8.67   New York Institute of Technology 2007—Solar Decathlon Project.
(Image courtesy of U.S. Department of Energy (DOE)/National Renewable Energy Laboratory (NREL))

- The house is constructed in three modular steel units.
- All the mechanical, electrical, and solar equipment is contained in the 270 ft^2 core.
- A geothermal well was chosen for an efficient and effective way of heating and cooling.
- Several shading devices to reduce solar penetration during the summer are installed on the south side, which reduces energy used for cooling the inside.
- A device displaying the amount of energy used and produced demonstrates easily how much energy people are using.
- Thirty four solar panels provide the majority of the energy used.
- The house has a rooftop pool with a fountain flowing down to a pond to be recycled and cooled.

PENN STATE 2007

Penn State was certainly up for the challenge of designing a Solar Decathlon home. In fact, they designed two of them. The competition home is called MorningStar Pennsylvania. Its final destination will be on the Penn State campus as a renewable energy research lab. The second solar house is MorningStar Montana and will find a home on the Northern Cheyenne Reservation as a visiting faculty guest house. The purpose of building both structures was to test their market concept.

The homes owe a lot to prefab materials, but some of the house is built on the site. As with other entries, the main mechanical center of the house, or the Technical Core, is prefabricated. Materials of opportunity (such as wood from a white oak tree that had fallen) will finish the structure.

Green materials

- Photovoltaic-powered LED lighting was installed in a curtain wall which changes color corresponding to weather.
- Reclaimed slate shingles along with some Pennsylvania bluestone finish the roof.
- The exterior cladding of the house is reclaimed Pennsylvania black slate. The cladding is a composite with other materials being: steel, hardwood, and milk bottles.

Green technologies

- The home contains an Energy Dashboard which monitors rates of energy consumption and of energy production. The display monitor allows the students to see how they are "spending their energy."
- The living area is characterized by its openness. The roof is on a slope to allow for clerestory windows.
- The exterior on the south is of recycled steel and white oak. Light and heat can be regulated with the sliding panels.
- The breezeway forms the seam or separation corridor between the technical core and the living area. As is the function of a breezeway, it channels air along its length.
- The roof garden acts as a water absorption and filtering device and is planted with colorful species that are good water absorption plants.

Figure 8.68   **Penn State 2007—Solar Decathlon Project.** *(Image courtesy of U.S. Department of Energy (DOE)/National Renewable Energy Laboratory (NREL))*

SANTA CLARA UNIVERSITY 2007

The Santa Clara University wants a visitor to their Ripple Home to see a mingling of traditional and contemporary found in California style. The term "ripple" was chosen because the students wanted their house to start a ripple effect of educating the public to solar power. Very typical of the state is the huge deck—600 ft^2 of it—with a wheelchair accessible ramp. The two pieces of the staggered home consist of the open living and kitchen areas and, in the north module, the bedroom. The bathroom is between the two areas. Bringing the inside of the house into the outdoors is another California tradition. The NanaWall, an 8-ft wall of insulated glass, can be a wall or a window. It folds away, allowing the living room to suddenly become part of the "fun in the sun" California lifestyle.

Green materials

- Plenty of wood is used in and around the home: such as, redwood planter boxes, teak patio furniture, and wine barrel stave chairs.
- Fiber cement siding composes the south and west walls, chosen because of its insulation properties and durability.
- Other walls are made of paper pulp treated with resin and baked in solid sheets; this material was chosen for its heat-transmitting properties.

Figure 8.69   Santa Clara University 2007—Solar Decathlon Project.
(Image courtesy of U.S. Department of Energy (DOE)/National Renewable Energy Laboratory (NREL))

- Fluorescent lighting is the main choice, controlled manually or through a computer interface.
- Bamboo cabinetry, handcrafted by the students, beautifies the house.

Green technologies

- A planted trellis shades the south side of the home.
- Thirty four photovoltaic panels power the house, and the battery reserve holds five days of reserve power.
- An integrated control system monitors and adjusts light, temperature, humidity, carbon dioxide levels, etc., equipped, of course, with overrides.
- The control system can be accessed with a cell phone if the owner is away from home.

TEAM MONTREAL 2007

Team Montreal wanted to convince people that solar power does not just work in warm climates. To achieve this goal they have started a Polar Revolution. As they see it, the most important aspect of their solar home was the building envelope, which must absorb and trap heat while still looking great. Their aesthetically-pleasing facade consists of wood siding; the front entrance is recessed, providing a buffer in a cold climate. A green wall covers the front facade with foliage except for the long narrow windows.

The floor plan is open, with the living room on the east with folding doors leading out to the front step. On the west is the study with a bedroom behind. This space also has matching folding doors that exit to the front step. The kitchen, dining room, mechanical room, and the bathroom are lined up along the north wall from east to west.

Green materials

- A steel frame is easy to assemble and disassemble.
- Soybean and recycled plastic are some of the materials found in the polyurethane insulation.
- Windows have automated shading to hold in heat.
- Fluorescent lighting is the main choice, controlled manually or through a computer interface.
- Windows are triple glazed and filled with argon.

Green technologies

- Walls are clipped on to the steel frame of the house.
- Forty photovoltaic panels on the roof are clipped to the structure in the same way the walls are clipped to the frame. The photovoltaic panels act as the only roofing material.
- The house boasts a green roof as well as a green wall, which add insulation value and trap rainwater.
- Two solar thermal collectors take care of hot water for radiant floors and domestic use.
- The "smart" energy system can search the web for the weather forecast, predict how much energy it will need, and recommend energy use amounts to prepare for the future.

Figure 8.70   **Team Montreal 2007—Solar Decathlon Project.** *(Image courtesy of U.S. Department of Energy (DOE)/National Renewable Energy Laboratory (NREL))*

TECHNISCHE UNIVERSITÄT DARMSTADT 2007

The Technische Universität Darmstadt team captured first place in the architectural part of the Solar Decathlon with their simple design. Some of the factors that influenced the judges' decision were the entry into the home (five folding German oak doors equipped with movable shutters), ability to circulate around the plan of the interior, accommodation of the technological portion of the house, adequate living space, and quality of materials composing the house.

Final plans for the house entail its removal to the University campus to be used as a solar power plant. The German government encourages solar power by guaranteeing a fairly generous amount of money to power fed into the grid.

Green materials

- Wide oak flooring.
- Vacuum-insulated walls in the second layer provide a longer lasting life span (above 15 years) and provide for higher energy retention in the home. The cost and production of these panels is expensive, but they effectively save energy and are very durable.
- The photovoltaic systems are installed not only on the roof, but are also successfully disguised in the moveable shutters that make up the six doors and all the rest of the façade.

Figure 8.71   Technische Universität Darmstadt 2007—Solar Decathlon Project.
(Image courtesy of U.S. Department of Energy (DOE)/National Renewable Energy Laboratory (NREL))

Green technologies

- This home was designed with three layers that control different functions. The first layer of wood and photovoltaic louvers provide shade and generate electricity.
- The second layer is a thermal envelope of opaque, vacuum-insulated walls on the east and west sides and floor-to-ceiling windows on the north and south sides.
- The third layer is the central core of the home in which all vertical technical installations and the private facilities are bundled.
- The house rests on a platform system, allowing for quick assembly. Under the platform are concealed the technological systems and storage space.
- The structure of the house is in three modules.
- Furniture can be folded into the platform floor, saving space.
- The building envelope consists of oak louvered frames, all equipped with photovoltaic panels.
- The flat roof was an interesting deviation from the other entries. It is designed with multiple layers with 10 times better insulation than a traditional roof.

TEXAS A&M UNIVERSITY 2007

The Texas A&M University brought the groHome concept to the contest. This is a concept that allows one to switch entire rooms around. One student states: "You could swap the position of the kitchen and the bath without a problem, buy an extra kitchen on eBay, or sell off a couple of rooms after the kids move out." This is possible through interconnected groWall units. Some of these even have the entire bath or kitchen utilities built into them. The home's flexibility will be an important asset over time.

Green materials

- The steel columns and beams form the skeleton on the inner layer of the house.
- A skin is attached to the skeleton made up of groWall units and SIPs.
- Cables about 2 ft away from the walls support photovoltaic panels, flower trellises, etc.
- Lightweight plastic panels cover the outside of the wall.
- This entry even had a hot tub.
- Racks of photovoltaics extend over the edge of the roof.
- The north side of the house holds evacuated tube solar collectors which provide hot water for space heating and domestic use.
- Lighting consists of thin, bendable light-emitting capacitors.
- A deck, pool with fish, and a "bat tower" completes the house.

Green technologies

- Lumber sunshades are used both to reduce solar gain and allow diffuse lighting. They also function as a strong architectural element. The rear sunshades are made of steel and are more an architectural element than a functioning sunshade.
- The house rests on steel grate footings.
- The core of the house, an actual trailer, holds the mechanical systems. The trailer rests on the foundations and is not attached to the house.

Figure 8.72   Texas A&M University 2007—Solar Decathlon Project. *(Image courtesy of U.S. Department of Energy (DOE)/National Renewable Energy Laboratory (NREL))*

■ The groWalls are attached to the core.
■ The porch is a separate module, as is the garage.

UNIVERSIDAD POLITÉCNICA DE MADRID 2007

Universidad Politecnica de Madrid brought an entry with a dramatic design. The roof is one of the most striking features, being steeply sloped and of a solid imposing form. Electrochromic windows, a landscape zone, a vegetable panel, and a living wall are some of the unique green features of this project. The house is designed in separate modules to combine the technology, to eliminate the heat gain, and to provide easy transportation. A large deck is located on the south side to expose its interior to the sun.

Green materials
■ Steel studs compose the framework.
■ A landscape zone, vegetable panel, and water sheet help to naturally regulate temperatures.
■ Windows are electrochromic, darkening or lightening as sun levels change.
■ The living wall on the outside highlights the house's interaction with nature.

Green technologies
■ Solar photovoltaic cells, installed on the mobile roof that rotates to follow the sun, produce power.

Figure 8.73   Universidad Politécnica de Madrid 2007—Solar Decathlon Project.
(Image courtesy of U.S. Department of Energy (DOE)/National Renewable Energy Laboratory (NREL))

- A double envelope on the house provides exceptional insulation.
- Phase changing gels are installed in the foundation, providing thermal control.
- The house is divided into three areas, one of which houses all the necessary technology along with bathroom and kitchen. The other two areas are open, for the main living space.
- The west side of the house has no openings, helping to eliminate heat gain during the hottest part of the day.

UNIVERSITY OF PUERTO RICO 2007

The University of Puerto Rico team found inspiration in the analysis of a single cell. Their design is an example of biomimicry, or the adoption of nature's designs for human construction. "The cell produces energy, recycles waste, adapts to changing conditions, functions independently, and communicates with other cells." The team aspired to design their home to fulfill the same functions. Since the nucleus is the essential element in the cell, the nucleus of their house holds the necessary components to make the home function. The students located the bathroom, battery room, laundry, and kitchen in the house nucleus. The rest of the house might be compared to the space surrounding the nucleus. A studio/bedroom contains a folding bed which increases the space available.

Figure 8.74   University of Puerto Rico 2007—Solar Decathlon Project.
(Image courtesy of U.S. Department of Energy (DOE)/National Renewable Energy Laboratory (NREL))

Like a cell, which makes use of its waste, the house also purifies water for reuse, uses a reflective pond to collect rainwater, and features a garden that grows some food.

Green materials

- Recycled wood for flooring and walls reduces waste and increases the lifetime and usability of the material.
- Ecoresin panels are used for the exterior. These are insulated, reflective, and made of recycled material. The panels come in almost every color and texture imaginable.
- Louvered screens provide dual performance: shade when windows are open as well as an array of different natural lighting options depending on the season or the wishes of the occupant.

Green technologies

- The Ecoresin panels help reflect the elements, increasing the lifetime of the siding.
- Solar panels are arranged in a pergola to create a roof that connects the battery room to the terrace so they also perform as a roof membrane.
- To better enable transportation, the house uses lightweight components and was assembled from two modules. The first half contains electrical equipment, while the second holds the water system.

UNIVERSITY OF COLORADO AT BOULDER 2007

Although the contest regulations demanded a maximum of 800 ft^2, the University of Colorado's project was designed as a full size (2100 ft^2) house with three equal pieces (700 ft^2 each). However, only one piece was presented and displayed at the competition. Most of the materials for the project were chosen from reclaimed and post-industrial recycled materials. Photovoltaic systems were integrated into the building along with a thermal collection system. Some of the primary green technologies included: smart glass, a prefabricated spine with a centralized HVAC, passive solar overhangs, and thermal storage tanks.

Green materials

- The use of recycled shipping containers promises a strong core.
- Clear ecoresin doors and wall panels are used as ecologically sustainable, nontoxic, and 40 percent post-industrial recycled material. The particular panels used are 40 times more durable than glass and are chemical resistant.
- Flooring that is a composite of linseed oil, wood flour, jute, pine resin, limestone, and environmentally friendly pigments was chosen.
- Recycled paper-based fiber composite countertops are used in the kitchen area.
- Cabinetry is made from laminated bamboo. The plywood is formaldehyde-free.
- Reclaimed redwood is used on the interior and deck.
- Photovoltaic panels are integrated into the building, along with a photovoltaic thermal collection system. The thermal system takes heat from the back of the photovoltaic array and transfers it to storage for a water-to-water heat pump.
- A battery bank holds reserve energy for four days of zero photovoltaic power.

Figure 8.75   University of Colorado at Boulder 2007—Solar Decathlon Project.
(Image courtesy of U.S. Department of Energy (DOE)/National Renewable Energy Laboratory (NREL))

Green technologies

- The larger house is designed to be built using two more shipping containers, economical for their prefabricated modular ease of construction.
- Using passive solar building overhangs blocks heat gain.
- A prefabricated spine with a centralized HVAC system localizes all energy use and keeps the core of the building warm or cool.
- Smart glass is a high-performance technology that electrically reflects heat and light or allows heat and light to enter, in response to a set voltage.
- A highly efficient water to water heat pump is used to heat both the air and the water as well as heat exchangers. The water retains the heat better and transfers energy quicker, using less energy, especially since the two are simultaneously heated.
- Thermal storage tanks with a two-day capacity retain water heat, minimizing the need for extra heating.

UNIVERSITY OF CINCINNATI 2007

The University of Cincinnati home has not just one sun-collecting southern wall; it has two. The first "wall" is actually a fence a few feet away from the facade. Its purpose is to hold 120 evacuated tube solar thermal collectors. As these collectors harvest the sun's energy, the fence also partially shades the courtyard. The second wall, all glass, is the south facade of the house, facing the courtyard. Thus, the house gets double benefits from the sun just from the south side.

Green materials

- The butterfly roof works as a shading device.
- Students created and assembled all of the furniture.
- Triple-paned low-E glass is used in the windows on the south.
- Daylighting is made even more efficient from clerestory windows all around the house.
- The exterior of the house has colorful tiles derived from recycled Formica plus reclaimed metal, positioned about three inches away from the actual walls, that act as a rainscreen by protecting the walls from moisture.
- A complicated "zipper" steel frame links sections of the house together.
- Rubber flooring is used throughout the interior.

Green technologies

- The fence on the south is actually a conveyance holding 120 evacuated tube solar thermal collectors. Filled with water, they provide heating as well as cooling, by means of an absorption chiller.
- The absorption chiller works like a typical A/C unit, but runs much more efficiently. It uses hot water from the evacuated tubes to activate an absorption cooling cycle.
- Modularity and prefabrication reduces on-site construction.
- A solar trellis has a 7.5-kW array.
- From the butterfly roof the gutter is sloped to the west, where a gray water collection tank is located.
- Heating, cooling, electrical, plumbing, dehumidification, and communication are housed in a compact easy access area.

Figure 8.76 University of Cincinnati 2007—Solar Decathlon Project.
(Image courtesy of U.S. Department of Energy (DOE)/National Renewable Energy Laboratory (NREL))

UNIVERSITY OF ILLINOIS AT URBANA-CHAMPAIGN 2007

University of Illinois at Urbana-Champaign (UIUC) house was constructed in three modules produced in a warehouse, and the students believe they have the ability to mass-produce these modules.

Their innovative solution to heating and cooling the house deserves mention. All of it is done through radiant heating located in the ceiling of the house.

The students believe their home, Elementhouse, will be appreciated by the ordinary consumer for its affordability. "It's the Volkswagen of homes," they remarked. All furniture is designed and customized by the students themselves.

Green materials

■ The cabinetry throughout the house is made from 100 percent recycled particle board.
■ Polyurethane foam provides the insulation in the wall panels.
■ Fluorescent lights and LED bulbs are used.

Green technologies

■ The cooling and heating of the building is contained and created by the radiant ceiling panels which in turn, release no harmful wastes into the environment. These panels easily warm and cool the spaces because of their location on the ceiling.
■ Large windows that are made of cellular glass allow the building to receive heat energy from the sun and also allow natural light to fill the space.
■ The 40 photovoltaic panels return so much energy to the project that it can power itself all year.

Figure 8.77   **UIUC 2007—Solar Decathlon Project.** *(Image courtesy of U.S. Department of Energy (DOE)/National Renewable Energy Laboratory (NREL))*

UNIVERSITY OF MARYLAND 2007

The students of the University of Maryland see their house as "the elegant marriage of biological knowledge and cutting-edge technology". The design for their home, LEAFhouse, was inspired by the leaf, nature's unique way of converting sunlight into energy. They saw the leaf as "nature's most efficient organism". Another feature connecting the house to the natural world is the skylights that cross the roof from east to west.

The home is as efficient as possible, with most of the energy coming from the sun. The team created a smart house system called SHAC (Smart House Adaptive Control) to bring comfort levels to a peak. "The network monitors humidity, temperature, light, and whether the doors are open or closed."

Green materials
- Exposed steel supports branch out in the living space.
- Many interior walls are translucent, moveable panels that transform static space.
- Large expanses of glass bring in natural light, which also passively heats the space.
- The siding of the home is a composite of wood, glass, vegetation, and corrugated metal.
- Spray foam insulation helps to reduce heating costs.
- A liquid dessicant waterfall on the living room wall not only controls humidity but is an aesthetic and unique feature of the house.
- A Murphy bed folds away and walls slide back, allowing bedroom space to become part of the living space.

Green technologies
- The photovoltaic power stem provides 100 percent of electrical energy required to operate the house.
- Solar water heating tubes, created by Apricus Solar Co., Ltd, provide all the hot water.
- Grey water is collected from the sink and washing machine; it is filtered and then stored for yard maintenance.
- Rainwater is collected from the roof, filtered by the vertical rain garden on the south wall, and used to irrigate the garden.
- A radiant floor heating system controls the temperature.
- Louvers over large sliding glass doors on the south can shade or light the interior.
- Maple countertops beautify the kitchen area.
- Skylights further brighten the interior.
- Windows and doors are oriented to take advantage of natural breezes, in turn naturally cooling the spaces.
- Nearly completely surrounded by a deck or plantings, the house embraces the outdoors.
- The bed can be folded up, the walls thrown out, and the space becomes part of the rest of the living area.

Figure 8.78   **University of Maryland 2007—Solar Decathlon Project.**
(Image courtesy of U.S. Department of Energy (DOE)/National Renewable Energy Laboratory (NREL))

UNIVERSITY OF MISSOURI-ROLLA 2007

After participating in the Solar Decathlon competition for three years, the University of Missouri-Rolla now has a campus "solar village" used for student housing and research. In 2007, the students built a home that is clean-lined, minimalist, and is completely in harmony with nature. The Missouri-Rolla team believes that a middle-class family should be able to afford it.

The home is a rectangular shape with an extending east bedroom, but the entry is a curved cutaway shape out of the southwest corner, that is designed to invite visitors inside. The curved entry leads one in to the combined living, dining, kitchen area, and down a corridor to the east bedroom. The kitchen has an island set at an angle that visually separates this space from the living space. The bathroom and mechanical systems are along the north wall.

Green materials

- The exterior of the house is composed of panels made of a product called PaperStone. This is postconsumer waste and composed primarily of paper and non-petroleum resin.
- Lyptus hardwood and eucalyptus wood is used throughout the interior. Lyptus hardwood is made from a rapidly renewable resource.

Figure 8.79   University of Missouri-Rolla 2007—Solar Decathlon Project.
(Image courtesy of U.S. Department of Energy (DOE)/National Renewable Energy Laboratory (NREL))

Green technologies

- A central control room houses mechanical systems and is the "brain of the house."
- Many interior furnishings fold away or have many uses, thus saving space or redesigning the space.
- An automation system controls windows, air-conditioning, lighting, etc.
- The south wall is of folding glass that can open up to let interior space become part of the outdoors.

UNIVERSITY OF TEXAS AT AUSTIN 2007

The University of Texas at Austin brought their BloomHouse ("it blooms like a rose under the sun; it's about a budding way of life") to the contest. And the owner can "bloom" in the outdoor hot tub in water heated from excess heat from the water system.

The mechanical portion of the house is located along the west wall—pumps, hot water tank, radiant floor water tank, and electrical closet. Just to the east of this portion are storage cupboards backed up to the mechanical area. A surprisingly large bathroom and kitchen occupy the north wall and the center of the house, respectively. The living area/work station is on the east quarter of the house.

Green materials

- Old style wooden shutters evoke a feeling of nostalgia, and are endlessly adaptable to heat and light.

Figure 8.80   University of Texas at Austin 2007—Solar Decathlon Project.
(Image courtesy of U.S. Department of Energy (DOE)/National Renewable Energy Laboratory (NREL))

- The photovoltaic system produces 7.6 kW and includes a roof brim (like the brim on a cap).
- Lightweight plastic panels cover the exterior and move with the passing wind.
- A hot tub adds a sybaritic touch.
- Rubber flooring is used for the mechanical core of the house.
- A huge deck with fiberglass decking was used.
- Caesarstone countertops are installed in the bathroom and kitchen.
- The main living area has hardwood floors.

Green technologies
- Natural cooling is used in preference to a great deal of air-conditioning.
- Radiant floor heating is used.

BIBLIOGRAPHY

Addington, M. and D. Schodek (2005). *Smart Materials and Technologies*. Burlington, MA, Architectural Press.

AIA (2009). COTE Mission. American Institute of Architects (AIA). Available at: http://www.aia.org.

Anastas, P. L. and J. B. Zimmerman (2003). "Through the 12 principles of green engineering." *Environmental Science and Technology* March **1**:95–10.

Anink, D., C. Boonstra and J. Mak (1996). *Handbook of Sustainable Building*. London, UK: James & James (Science Publishers) Limited.

Anneling, R. (1998). "The P-mark system for prefabricated houses in Sweden." Swedish National Testing and Research Institute CADDET (Centre for Analysis and Dissemination of Demonstrated Energy Technologies) Newsletter (1):20–22.

ANRC (2005). Valuing Ecosystem Services: Toward Better Environmental Decision-Making Committee on Assessing and Valuing the Services of Aquatic National Research Council. National Academic Press.

Architecture2030 (2009). "The 2030 Challenge." Available at: http://www.architecture2030.org.

Baillie, C., Ed. (2004). *Green Composites*. Cambridge, GB, Woodhead Publishing Limited.

Barrow, C. J. (1995). "Sustainable development: Concept, value and practice." *Third World Planning Review* **17**(4):369–386.

Barsoum, M. (1997). *Fundamentals of Ceramics*. New York, NY: McGraw-Hill.

Basalla, G. (1989). *The Evolution of Technology*. New York, NY: Cambridge University Press.

Beatley, T. and K. Manning (1997). *The Ecology of Place: Planning for Environment, Economy, and Community*. Washington, DC: Island Press.

Benestad, R. E. (2006). *Solar Activity and Earth's Climate*. New York, NY: Springer.

Berge, B. (2000). *The Ecology of Building Materials*. Oxford; Boston, MA: Architectural Press.

Berger, L. (2002). "Climate: An exceptionally long interglacial ahead?" *Science* **297**(5585):1287–1288.

Berinstein, P. (2001). *Alternative Energy*. Westport, CT: Oryx Press.

Bersch, J. (2009). *Cellulose: Cellulose Products and Rubber Substitutes*. Knowledge Publications.

Blacksmith_Institute (2008). *World's Worst Pollution Problems: The Top Ten of the Toxic Twenty*. New York, NY: Blacksmith Institute, Green Cross Switzerland.

Bridgman, P. W. (1980). *The Logic of Modern Physics*. New York, NY: Ayer.

Bridgwater, A., Ed. (2001). *Progress in Thermochemical Biomass Conversion*. Oxford; Malden, MA: Wiley-Blackwell.

Bridgwater, A., Ed. (2003). *Pyrolysis and Gasification of Biomass and Waste*. Newbury Perks, UK: CPL Press.

Bridgwater, A., Ed. (2008). *Advances in Thermochemical Biomass Conversion*. New York, NY: Springer.

Brinkhoff, T. (2009). The Principal Agglomerations of the World. Available at: http://www.citypopulation.de/ world/Agglomerations.html.

Broadbent, G. and C. A. Brebbia, Eds. (2006). *Eco-Architecture: Harmonisation Between Architecture and Nature*. Southampton, UK; Boston, MA: WIT Press.

Broswimmer, F. (2002). *Ecocide: A Short History of Mass Extinction of Species*. London; Sterling, VA: Pluto Press.

Brown, L. (2001). *Eco-Economy*. New York, NY: W. W. Norton.

Brownell, B. (2006). *Transmaterial: A Catalogue of Materials that Redefine Our Physical Environment*. New York, NY: Princeton Architectural Press.

Bube, R. H. (1998). *Photovoltaic Materials*. London: Imperial College Press; River Edge, NJ: Distributed by World Scientific Publishing Company.

Buchs, W., Ed. (2003). *Biotic Indicators for Biodiversity and Sustainable Agriculture*. Amsterdam; New York: Elsevier.

Bullen, D. (2006). "Building performance: Past, present and future." *AIA Journal of Architecture*. October 2006(1): 4–5.

Calow, P. P., Ed. (2008). *Evolutionary Physiological Ecology*. New York, NY: Cambridge University Press.

Caro, T., Ed. (1998). *Behavioral Ecology and Conservation Biology*. New York, NY: Oxford University Press.

Carslaw, K. S., R. G. Harrison, and J. Kirkby (2002). "Atmospheric science: Cosmic rays, clouds, and climate." *Science* (298):1732–1737.

Carson, R. (1962). *Silent Spring*. Cambridge, MA: Houghton Mifflin.

Carter, N. (2001). *The Politics of the Environment: Ideas, Activism, Policy*. New York, NY: Cambridge University Press.

Chang, K. F, C. M. Chiang, and P. C. Chou (2007). "Adapting aspects of GBTool 2005—Searching for suitability in Taiwan." *Building and Environment* **42**(1):310–316.

Chapin, S., H. A. Mooney, and M. C. Chapin (2004). *Principles of Terrestrial Ecosystem Ecology*. New York, NY: Springer.

Chawla, K. K. (2003). *Ceramic Matrix Composites*. Norwell, MA: Kluwer Academic Publishers.

Chevalier, J.-M. (2007). "Energy Economics and Energy Econometrics." *The Econometrics of Energy Systems*. J. H. Keppler, R. Bourbonnais, and J. H. Keppler, Eds. New York, NY: Palgrave Macmillan: Introduction, xiii-xxv.

Chew, M. Y. L. and S. Das (2004). "Building grading systems: a review of the state-of-the-art." *Journal of Architectural Engineering* **10**(3):80–87.

Christopherson, R. W. (1997). *Geosystems: An Introduction to Physical Geography*. Upper Saddle River, NJ: Prentice Hall.

Clark, T., R. Reading, and A. Clarke, Eds. (1994). *Endangered Species Recovery: Finding the Lessons, Improving the Process*. Washington, D.C.: Island Press.

Cowie, J. (2007). *Climate Change: Biological and Human Aspects*. New York, NY: Cambridge University Press.

Crawley, D. and I. Aho (1999). "Building environmental assessment methods: Applications and development trends." *Building Research and Information* **27**(4–5):300–308.

CSA (2001). S478-95 (Reaffirmed 2001), Guideline in Durability in Buildings, Canadian Standards Association (CSA).

Daniel, I. M. and O. Ishai (2006). *Engineering Mechanics of Composite Materials*. New York, NY: Oxford University Press, 2nd Edition.

Davis, M. (2006). *Planet of Slums*. New York, NY: Verso.

Deb, S. K. (1973). "Opt. and photoel. prop. of C.C. in thin films of tungsten oxide." *Philosophical Magazine*. 27: 801–822.

Demkin, J. A., Ed. (1996). *Environmental Resource Guide*. Hoboken, NJ: Wiley.

DOE (2006). *Biomass Energy Databook*. U.S. Department of Energy and Energy Information Administration (EIA).

DOE (2008). *International Energy Outlook—2008*. U.S. Department of Energy and Energy Information Administration (EIA).

Dorf, R. C. (2001). *Technology, Humans, and Society: Toward a Sustainable World*. Amsterdam, Holland; New York, NY: Elsevier.

Dresner, S. (2002). *The Principles of Sustainability*. London, UK: Earthscan Publications.

Duerig, T. W., Ed. (1990). *Engineering Aspects of Shape Memory Alloys*. London, UK: Butterworth-Heinemann.

Duffie, J. A. and W. A. Beckman (2006). *Solar Engineering of Thermal Processes*. Hoboken, NJ: Wiley.

Dugatkin, L. A., Ed. (2001). *Model Systems in Behavioral Ecology: Integrating Conceptual, Theoretical, and Empirical Approaches*. Princeton, NJ: Princeton University Press.

Earth Policy Institute Resources on Carbon Emissions (www.earth-policy.org/Indicators/CO2)

Earth Systems Research Laboratory (ESRL)/National Oceanic and Atmospheric Administration (NOAA) (www.esrl.noaa.gov/gmd/ccgg/trends)

Edwards, A. R. (2005). *The Sustainability Revolution*. Gabriola, BC: New Society Publishers.

Ehrenfeld, J. (2008). *Sustainability by Design: A Subversive Strategy for Transforming Our Consumer Culture*. New Haven, CT: Yale University Press.

Ehrlich, P. R. (1968). *The Population Bomb*. New York, NY: Ballantine Books.

Ehrlich, P. R. (1985). *Extinction: The Causes and Consequences of the Disappearance of Species*. New York, NY: Ballantine Books.

EPA (2008). Heating and Cooling Efficiency of Geothermal Heat Pumps. Energy Efficiency and Renewable Energy. U.S. DoE, Environmental Protection Agency (EPA).

ESRL (2009). *MLO Carbon Dioxide Trends.* Earth System Research Laboratory, Mauna Loa Observatory.

FAO (2001). *Global Forest Resources Assessment 2000.* Rome; Food and Agriculature Organization of the United Nations: p 482.

Farrelly, D. (1984). *The Book of Bamboo: A Comprehensive Guide to This Remarkable Plant, Its Uses, and Its History.* San Francisco, CA: Sierra Club Books.

Femandez, J. E. (2002). "Flax Fiber Reinforced Concrete—A Natural Fiber Biocomposite for Sustainable Building Materials." *High Performance Structures and Composites.* C. A. Brebbia and W. P. D. Wilde. Eds. Computational Mechanics, Inc.

Fernandez, J. E. (2005). *Material Architecture: Emergent Materials for Innovative Buildings and Ecological Construction.* Burlington, MA: Architectural Press.

Fowler, K. M. and E. M. Rauch (2006). "Sustainable building rating systems summary," Pacific Northwest National Laboratory, DoE.

Fox, C. W., D. A. Roff, and D. J. Fairbairn, Eds. (2001). *Evolutionary Ecology: Concepts and Case Studies.* New York, NY: Oxford University Press.

Fox, S. (1986). *American Conservation Movement: John Muir And His Legacy* Madison, WI: University of Wisconsin Press.

Francis, M. (2007). *Herbert Spencer and the Invention of Modern Life.* Newcastle, UK: Acumen Publishing.

Fraunhofer. Available at: http://www.iap.fraunhofer.de.

Frodin, D. G. (2001). *Guide to Standard Floras of the World.* New York, NY: Cambridge University Press.

Gaston, K. J. and J. I. Spicer (2004). *Biodiversity: An Introduction.* Oxford, Malden, MA: Wiley-Blackwell, 2nd Edition.

Gimpel, J. (1976). *The Medieval Machine.* New York, NY: Holt, Rinehart and Winston.

Gimpel, J. (1979). "Environmental Pollution in the Middle Ages." *Technology and Change.* J. Burke and M. Eakin, Eds. New York, NY: Holt, Rinehart and Winston: 135–140.

GISS (2009). GISS Surface Temperature Analysis, Goddard Institute for Space Studies (GISS). Available at: http://data.giss.nasa.gov/gistemp/graphs.

Gleick, P. H., H. Cooley, et al. (2007). *The World's Water 2006–2007*: *The Biennial Report on Freshwater Resources.* Washington, DC: Island Press.

Gockel, D. (1994). "Building decommission requires environmental review." Real Estate Weekly.

Golley, F. B. (1996). *A History of the Ecosystem Concept in Ecology: More than the Sum of the Parts.* New Haven, CT: Yale University Press.

Gorham, E. (2006). "Ecosystem." *Environmental Encyclopedia.* M. Bortman, P. Brimblecombe, and M. A. Cunningham, Eds. Detroit, MI, Gale. **1:**426–428.

Goudie, A. (2006). *The Human Impact on the Natural Environment.* Oxford, Malden, MA: Blackwell Publishing, 6th Edition.

Gowri, K. (2004). "Green building rating systems: An overview." *American Society of Heating, Refrigerating, and Air-Conditioning Engineers (ASHRAE)* November 2004:56–59.

Graham, P. (2002). *Building Ecology.* Malden, MA; Oxford Blackwell Publishing.

Granqvist, C. G. (1995). *Handbook of inorganic electrochromic materials.* Amsterdam; New York, NY: Elsevier.

Hanan, J. J. (1997). *Greenhouses: Advanced Technology for Protected Horticulture*. Boca Raton, FL: CRC Press.

Hannah, L., D. Lohse, C. Hutchinson, J. L. Carr, and A. Lankerani (1994). "A preliminary survey of inventory of human disturbance of world ecosystems." *Ambio* (23):246–250.

Hannah, L., J. L. Carr, and A. Lankerani (1995). "Human disturbance and natural habitat: A biome level analysis of a global data set." *Biodiversity and Conservation* (4):128–155.

Hinrichsen, D. and B. Robey (2000). *Population and the Environment: The Global Challenge, Center for Communication Programs*. Baltimore, MD: School of Public Health, Johns Hopkins University.

Hoadley, B. (2000). *Understanding Wood: A Craftsman's Guide to Wood Technology*. Newtown, CT: Taunton Press, Revised Edition.

Horvat, M. and P. Fazio (2005). "Comparative review of existing certification programs and performance assessment tools for residential buildings." *Architectural Science Review* **48**(1):69–80.

Hottel, H. C. and J. B. Howard (1971). *New Energy Technology—Some Facts and Assessments*. Cambridge, MA: MIT Press.

Houghton, J. T., Y. Ding, D. J. Griggs, M. Noguer, P. J. v. d. Linden, and D. Xiaosu, Eds. (2001). *Climate Change 2001: The Scientific Basis*. New York, NY: Cambridge University Press.

Hoyt, D. V. and K. H. Schatten (1997). *The Role of the Sun in Climate Change*. New York, NY: Oxford University Press.

Hubbard, G. (2009). *Green Building Rating Systems: A Comparison of LEED and China's 3 Star*. Portland, ME: Fore Solutions: Consultants for High Performance Buildings.

Hubert, M. K. (1971). "The Energy Resources of the Earth." *Scientific American* (225):60–84.

Hudson, B. J. (1979). "Coastal Land Reclamation with Special Reference to Hong Kong." *Reclamation Review* (2):3–16.

Hughes, D. (1975). *Ecology in Ancient Civilizations*. Albuquerque, NM: University of New Mexico Press.

Hutson, S., N. Barber, J. Kenny, K. Linsey, D. Lumia, and M. Maupin (2004). *Estimated Use of Water in the United States in 2000*. Reston, VA: U.S. Geological Survey.

IEA (2005). Key World Energy Statistics. Paris, International Energy Agency (IEA).

IPCC (2007). "Climate Change 2007: Mitigation of Climate Change." *Contribution of Working Group III to the Fourth Assessment Report of the Intergovernmental Panel on Climate Change, 2007*. M. B, D. O. R., Bosch P. R., D. R. and M. L. A. Cambridge, UK: Intergovernmental Panel of Climate Change (IPCC).

Isaacson, W. (2004). *Benjamin Franklin: An American Life*. New York, NY: Simon & Schuster.

IUCN (2008). 2008 International Union for Conservation of Nature and Natural Resources (IUCN) Red List of Threatened Species. Available at: http://www.iucnredlist.org.

Jackson, T. (1996). *Material Concerns: Pollution, Profit and Quality of Life*. London: Routledge.

Jones, S., R. D. Martin, and D. R. Pilbeam, Eds. (1994). *The Cambridge Encyclopedia of Human Evolution*. New York, NY: Cambridge University Press.

Kambezidis, H. and C. Gueymard (2004). *Solar Radiation and Daylight Models*. Oxford; Burlington, MA: Elsevier Butterworth-Heinemann, 2nd Edition.

Kaplan, D. L., Ed. (1998). *Biopolymers from Renewable Resources*. New York, NY: Springer.

Karlessia, T., M. Santamourisa, K. Apostolakisb, A. Synnefaa, and I. Livada (2009). "Development and testing of thermochromic coatings for buildings and urban structures." Solar Energy **83**(4):538–551.

Kibert, C. J. (2007). *Sustainable Construction: Green Building Design and Delivery*. Hoboken, NJ: Wiley.

Klass, D. L. (1998). *Biomass for Renewable Energy, Fuels, and Chemicals*. San Diego, CA: Academic Press.

Koch, G. W. and H. A. Mooney (1995). *Physiological Ecology: Carbon Dioxide and Terrestrial Ecosystems*. San Diego, CA: Academic Press.

Kondratev, K. Y. and I. Galindo (1997). *Volcanic Activity and Climate*. Hampton, VA: A. Deepak Publishing.

Krapivin, V. F. and C. A. Varotsos (2008). *Biogeochemical Cycles in Globalizatoin and Sustainable Development*. New York, NY: Springer.

Krebs, J. R. and N. B. Davies, Eds. (1997*). Behavioural Ecology: An Evolutionary Approach*. Oxford; Malden, MA: Wiley-Blackwell.

Kuang, Y.-c. and J.-p. Ou (2007). "Passive smart self-repairing concrete beams by using shape memory alloy wires and fibers containing adhesives." *Journal of Central South University of Technology*. **15**(2008):411–417.

Lacinski, P. (2000). *Serious Straw Bale: A Home Construction Guide for All Climates*. White River Junction, VT: Chelsea Green Publishing Company.

Lagoudas, D., Ed. (2008). *Shape Memory Alloys*. New York, NY: Springer.

Lang, J. (1994). *Urban Design: The American Experience*. New York, NY: Van Nostrand Reinhold.

Lang, J., C. Burnette, M. Walter, and D. Vachon, Eds. (1974). *Designing for Human Behavior*. Stroudsburg, PA: Dowden, Hutchingon & Ross, Inc.

Lawrence, R. J. (2006). "Learning from the Vernacular." *Vernacular Architecture in the Twenty-First Century*. L. Asquith and M. Vellinga, Eds. New York: Taylor & Francis: 111–127.

Lee, W. L. and J. Burnett (2006). "Customization of GBTool in Hong Kong." *Building and Environment* **41**(12):1831–1846.

Leopold, A. (1949). *Sand County Almanac*. New York, NY: Oxford University Press.

Liu, Y., D. Prasad, J. Li, Y. Fu and J. Liu (2006). "Developing regionally specific environmental building tools for China." *Building Research and Information* **34**(4):372–386.

Lockwood, J. L. and M. L. McKinney (2001). *Biotic Homogenization*. New York, NY: Springer.

Loo, S. v. and J. Koppejan, Eds. (2008). *The Handbook of Biomass Combustion and Co-firing*. London, UK: Earthscan Publications Ltd.

Lopez, R. and M. A. Toman, Eds. (2006). *Economic Development and Environmental Sustainability: New Policy Options*. New York, NY: Oxford University Press.

Lovejoy, T. E. and L. Hannah, Eds. (2005). *Climate Change and Biodiversity*. New Haven, CT: Yale University Press.

Manfra, L. (2006). Living Breathing Buildings: Envisioning architecture that performs like natural organisms. *Metropolis*. 5:52–54.

Marsh, G. P. (2006). *Man and Nature, Scholarly Publishing Office*. Ann Arbor, MI: Scholarly Publishing Office, University of Michigan Library, Digitally Reprinted.

Marshall, R. (2006). *How to Build Your Own Greenhouse*. North Adams, MA: Storey Publishing.

Matthews, F. L. and R. D. Rawlings (1994). *Composite Materials: Engineering and Science*. London, UK; New York, NY: Chapman & Hall.

Mauna Loa Observatory (2009), Scripps Institute of Oceanography (SIO). Available at: http://www. mlo.noaa.gov/home.html.

McClellan, J. E. and H. Dorn (2006). *Science and Technology in World History*. Baltimore, MD: John Hopkins University Press.

McCullagh, J. C. (1978). *The Solar Greenhouse Book*. Emmaus, PA: Rodale Press.

McHenry, P. G. (1989). *Adobe and Rammed Earth Buildings: Design and Construction*. Tucson, AZ: University of Arizona Press.

McMasters, K. (2006). The Mother of Invention: This young Brooklyn firm's process – necessarily fast and cheap – is quickly earning them a reputation for ingenuity. *Metropolis*. **12**:68–70.

Merchant, C. (2005). *The Columbia Guide to American Environmental History*. New York, NY: Columbia University Press.

Meyer, W. B. and B. L. Turner (1994). *Changes in Land Use and Land Cover: A Global Perspective*. New York, NY: Cambridge University Press.

Millennium Ecosystem Assessment (2005). *Ecosystems and Human Well-being: Current State and Trends*. Washington, DC: Island Press.

Millennium Ecosystem Assessment (2005). *Ecosystems and Human Well-being: Synthesis*. Washington, DC: Island Press.

Millennium Ecosystem Assessment (2005). *Living Beyond our Means: Natural Assets and Human Well-Being: Statement from the Board*. Washington, DC: World Resources Institute.

Montgomery, D. R. (2008). *Dirt: The Erosion of Civilizations*. Berkeley, CA: University of California Press.

Moo-Young, M., J. Lamptey, B. Glick, and H. Bungay (1987). *Biomass Conversion Technology: Principles and Practice*. New York, NY: Pergamon Press.

Moro, A. (2004). *"Rating systems in Italy. iiSBE (International Initiative for a Sustainable Built Environment)." Advanced Buildings News 2*. March 5.

Murphy, P. (2008). *Plan C: Community Survival Strategies*. Gabriola Island, BC: New Society Publishers.

Naar, J. S. (1976). *Design for a Limited Planet*. New York, NY: Ballantine Books.

NABERS (2005). National Australian Built Environment Rating System New South Wales Government (NSW). Department of Environment and Climate Change, Australia.

Nachmias, C. and D. Nachmias (1996). *Research Methods in the Social Sciences*. New York, NY: St. Martin's Press, 5th Edition.

Naiman, R., R. E. Bilby, and S. Kantor, Eds. (1998). *River Ecology and Management*. New York, NY: Springer.

Nash, R. (2001). *Wilderness and the American Mind*. New Haven, CT: Yale University Press, 4th Edition.

Newman, O. (1972). *Defensible Space: Crime Prevention Through Urban Design*. New York, NY: Macmillan Publishing Company.

Newman, P. and I. Jennings (2008). *Cities as Sustainable Ecosystems*. Washington, D.C.: Island Press.

Nie, M. S., Y. G. Qin, Y. Jiang, and T. X. Song (2002). *China Eco-Housing Technology Assessment Handbook (CHETAH)*. Beijing, China: Architecture and Building Press.

Nielsen, R. (2006). *The Little Green Handbook: Seven Trends Shaping the Future of Our Planet*. New York, NY: Picador.

Odum, E. (1971). *Fundamentals of Ecology*. New York, NY: Saunders.

OECD (1992). Energy Statistics and Balances of Non-OECD Countries 1991–1992. Paris, France: Organisation for Economic Co-Operation and Development (OECD).

OECD (2003). Environmentally Sustainable Buildings. Paris, France: Organisation for Economic Co-Operation and Development (OECD).

Ohta, H. (2007). "Thermoelectrics based on strontium titanate." *Materials Today* **10**(10):44–49.

Olson, R. and D. Rejeski, Eds. (2004). *Environmentalism and the Technologies of Tomorrow: Shaping The Next Industrial Revolution*. Washington, D.C.: Island Press.

Opie, J. (1998). *Nature's Nation: An Environmental History of the United States*. New York, NY: Wadsworth Publishing.

Ostfeld, R. S., F. Keesing, and Eviner (2008). *Infectious Disease Ecology: Effects of Ecosystems on Disease and of Disease on Ecosystems*. Princeton, NJ: Princeton University Press.

Pastor, J. (2008). *Mathematical Ecology of Populations and Ecosystems*. Hoboken, NJ: Wiley-Blackwell.

Patrick, R. L. (1988). *Treatise on Adhesion and Adhesives (Volume 6)*. New York, NY: CRC.

Peuser, F. A., K.-H. Remmers, and M. Schnauss (2002). *Solar Thermal Systems: Successful Planning and Construction*. London, UK: Earthscan Publications Ltd.

Pianka, E. R. (1999). *Evolutionary Ecology*. San Francisco, CA: Benjamin Cummings Publishing, 6th Edition.

Pimentel, D., M. Tort, L. D'Anna, A. Krawic, J. Berger, J. Rossman, F. Mugo, et al. (1998). "Ecology of Increasing Disease: Population growth and environmental degradation." *Bioscience* **48**(10).

Ponting, C. (2007). *A New Green History of the World: The Environment and the Collapse of Great Civilizations*. New York, NY: Penguin Books, Revised Edition.

Porter, T. (2007). *Wood: Identification & Use*. London, UK: Guild of Master Craftsman Publications.

Prüss-Üstün, A., R. Bos, F. Gore, and J. Bartram (2008). Safer Water, Better Health: Costs, benefits and sustainability of interventions to protect and promote health. Geneva, Switzerland: World Health Organization (WHO).

REN21 (2009). *Renewables: Global Status Report, 2009 Update*. Renewable Energy Policy Network. Available at: http://www.ren21.net/pdf/RE_GSR_2009_Update.pdf.

Reupke, P., G. Sarwar, and A. S. Tariq (1994). *Biomass Combustion Systems: A Guide for Monitoring & Efficient Operation*. New York, NY: Hyperion Books.

Roodman, D. M. and N. Lenssen (1995). *A Building Revolution: How Ecology and Health Concerns are Transforming Construction*. World Watch Paper.

Rose, C. (2004). "Designing for Composites: Traditional and Future views." *Green Composites* C. Baillie. Cambridge, UK: Woodhead Publishing Limited: 9-22.

Salby, M. and P. Callaghan (2004). "Evidence of the Solar Cycle in the General Circulation of the Stratosphere." *Journal of Climate* (17):34–46.

Saltzman, B. (2001). *Dynamical Paleoclimatology: Generalized Theory of Global Climate Change*. San Diego, CA: Academic Press.

Sandler, K. (2003). "Survey Statistics: Analyzing What's Recyclable in C&D Debris." *Biocycle* November 2003:51–54.

Sark, W. v., M. Patel, A. Faaij, and M. Hoogwijk (2006). "The Potential of Renewables as a Feedstock for Chemistry and Energy." *Renewables-Based Technology*. J. Dewulf and H. V. Langenhove (Eds). Hoboken, NJ: Wiley.

Sassi, P. (2006). *Strategies for Sustainable Architecture*. London, UK; New York, NY: Taylor & Francis.

Saxena, A. (2003). *Control of Biotic and Abiotic Factors in Aquaculture*. Delhi, India: Daya Publishing House.

Scarre, C. (1993). *Smithsonian Timelines of the Ancient World*. New York, NY: Dorling Kindersley.

Schaffer, J. P. (2000). *Yosemite National Park*. Berkeley, CA: Wilderness Press.

Schwartz, M. M. (1997). *Composite Materials (Volume 1)*. Upper Saddle River, NJ: Prentice Hall.

Schwartz, M. M., Ed. (2008). *Smart Materials*. Boca Raton, FL: CRC Press.

Shanks, R. A. (2004). "Alternative Solutions: Recyclable Synthetic Fibre-Thermoplastic Composites." *Green Composites*, C. Baillie (Eds). Cambridge, UK: Woodhead Publishing Limited: 100-122.

Smith, P. (2006). *Sustainability at the Cutting Edge: Emerging Technologies for low energy buildings*. Oxford, UK; Burlington, MA: Architectural Press, 2nd Edition.

Song, C., A. M. Gaffney, and K. Fujimoto, Eds. (2002). CO_2 Conversion and Utilization. Oxford, UK; Washington, DC: American Chemical Society Publication.

Spencer, H. (1897). *First Principles*. New York, NY: D. Appleton and Company.

Spiegel, R. and D. Meadows (2006). *Green Building Materials: A Guide to Product Selection and Specification*. Hoboken, NJ: John Wiley & Sons, 2nd Edition.

Steele, J. (2005). *Ecological Architecture: A Critical History*. New York, NY: Thames & Hudson.

Steinfeld, C. (2006). "Sustainable Architecture." *Environmental Encyclopedia*. M. Bortman, P. Brimblecombe, and M. A. Cunningham, Eds. Detroit, MI, Gale. **1**:426–428. 3rd Edition.

Stevens, E. S. (2002). *Green Plastics*. Princeton, NJ: Princeton University Press.

Stringer, C. and P. Andrews (2005). *The Complete World of Human Evolution*. London, UK; New York, NY: Thames & Hudson.

Svensmark, H. (2007). *The Chilling Stars: The New Theory of Climate Change*. London, UK: Totem Books.

Talshir, G. (2002). *The Political Ideology of Green Parties: From the Politics of Nature to Redefining the Nature of Politics*. Hampshire, UK: Palgrave Macmillan.

Tansley, A. G. (1935). "The Use and Abuse of Vegetational Concepts and Terms." *Ecology* **16**:284–307.

Teeple, J. B. (2006). *Timelines of World History*. New York, NY: Dorling Kindersley.

Teich, A. H. (2003). *Technology and the Future*. Florence, KY: Wadsworth Publishing, 9th Edition.

Tester, J. W., E. M. Drake, M. Golay, M. Driscoll, and W. Peters (2005). *Sustainable Energy*. Cambridge, MA: MIT Press.

Thoreau, H. D. (1854). *Walden; or, Life in the Woods*. Boston, MA: Ticknor and Fields.

Trusty, W. B. (2000). "Introducing an assessment tool classification system." *Royal Architectural Institute of Canada Advanced Building Newsletter* (25):18–19.

U.N. Population Division (2006). *World Population Prospects: The 2006 Revision*. United Nations.

U.S. Census Bureau (2008). World POPClock Projection. I. P. Center, U.S. Census Bureau.

UN (1987). Our Common Future Development and International Co-operation: Environment. Report of the World Commission on Environment and Development.

UN (1992). *Rio Declaration on Environment and Development*. Report of the United Nations Conference on Environment and Development. United Nations Department of Economic and Social Affairs (DESA).

UN-HABITAT Urban Info (2009). United Nations Urban Settlements Program. Available at: http://www.devinfo.info/urbaninfo/home.aspx.

UNEP (2006). Buildings and Climate Change: Status, Challenges and Opportunities. p. 87.

UNEP/GRID (2008) Arendal Maps and Graphics Library. Available at: http://maps.grida.no/go/graphic/conversion-of-terrestrial-biomes.

USDE (2008). *Building Energy Databook*. U.S. Department of Energy and Energy Information Administration (EIA): Tables 1.1–1.6.

Vallero, D. A. and C. Brasier (2008). *Sustainable Design: The Science of Sustainability and Green Engineering*. Hoboken, NJ: John Wiley & Sons.

Vita-Finzi, C. (2008). *The Sun: A User's Manual*. New York, NY: Springer.

Vogtlander, J. G. (2001). *The Model of the Eco-Costs/Value Ratio: A New LCA-based Decision Support Tool*. Delft, The Netherlands: Delft University of Technology.

Waltner-Toews, D., J. J. Kay, and N.-M. E. Lister (2008). *The Ecosystem Approach: Complexity, Uncertainty, and Managing for Sustainability*. New York, NY: Columbia University Press.

Wang, Z. L. (2006). "Nanowires." *Nanotechnology*. L. E. Foster, Ed. Prentice Hall: 154–160.

Wardle, D. A. (2002). *Communities and Ecosystems: Linking the Aboveground and Belowground Components*. Princeton, NJ: Princeton University Press.

Warren, L. S., Ed. (2003). *American Environmental History*. Malden, MA: Wiley-Blackwell.

Wayman, M. and S. Parekh (1991). *Biotechnology of Biomass Conversion: Fuels and Chemicals from Renewable Resources*. Hoboken, NJ: Wiley.

WB (2007). *2007 World Development Indicators: Statistics*. Washington, DC: World Bank. Available at: http://web.worldbank.org/data.

WB (2008). *Water Resources Management: Groundwater*. Washington, DC: World Bank. Available at: http://web.worldbank.org/data.

WCED (1987). *Our Common Future*. New York, NY: Oxford University Press.

Wereko-Brobby, C. Y. and E. B. Hagan (1996). *Biomass Conversion and Technology*. Hoboken, NJ: Wiley.

Wheeler, D. R., U. Deichmann, K. D. Pandey, K. E. Hamilton, B. Ostro, and K. Bolt (2007). Air Pollution in World Cities database: Air Pollution Data by Country, World Bank. Available at: http://go.worldbank.org/3RDFO7T6M0.

WHO (1992). *Our Planet, Our Health*. Report of the WHO commission on health and environment. Geneva: World Health Organization (WHO).

WHO (1995). *Bridging the Gaps*. Geneva: World Health Organization (WHO).

Williams, D. (2007). *Sustainable Design: Ecology, Architecture, and Planning*. Hoboken, NJ: Wiley.

Wilson, M., K. Kannangara, G. Smith, M. Simmons, and B. Raguse (2002). *NanoTechnology: Basic Science and Emerging Technologies*. Boca Raton, FL: Chapman&Hall/CRC.

Wines, J. (2000). *Green Architecture*. Koln, Germany; New York, NY: Taschen.

Wool, R. and S. Sun (2005). *Bio-Based Polymers and Composites*. San Diego, CA: Academic Press.

WWF (2008). *Living Planet Report 2008*. Gland, Switzerland: World Wide Fund for Nature (WWF) and Global Footprint Network.

Yu, L. (2008). *Biodegradable Polymer Blends and Composites from Renewable Resources*. Hoboken, NJ: Wiley.

INDEX

Verification of Sustainable Product Attributes

The growth of the number of new products and innovative approaches that are introduced into the building construction industry on a continuing basis has been phenomenal. In recent years, a large number of such new products or practices claim to be "green" or consistent with and promoting the goals of sustainable construction. Green building construction and green products are on a steep rise. The value of green building construction is projected to increase to $60 billion by 2010 (Source: McGraw-Hill Construction, 2008—Key Trends in the European and U.S. Construction Marketplace: Smart Market Report. U.S. Green Building Council: Green Building by the Numbers, November 2008) and the green building products market is projected to be worth $30-$40 billion annually by 2010 (Source: Green Building Alliance, 2006—Green Building Products: Positioning Southwestern Pennsylvania as the U.S. Manufacturing Center. U.S. Green Building Council). This issue becomes more complicated considering the global market and the fact that many products manufactured in one country find their way to multiple countries around the globe. Because it is rather impossible for designers, contractors or code officials to verify the credibility of each and every such claim regarding sustainable attributes, it is of paramount importance for everyone who is impacted by buildings, i.e. general population, as well as all involved in the construction industry, to be able to have a reliable source of verification of such claims. This requires a methodology and a standardized process by which to evaluate the degree of "greenness" and sustainable attributes of construction materials, elements and assemblies to result in buildings' green performance.

The most credible program available to address the issue of green verification is a program established by the International Code Council Evaluation Service (ICC-ES) known as the Sustainable Attributes Verification and Evaluation™ (SAVE™). The ICC-ES SAVE™ Program provides independent verification of manufacturers' claims about the sustainable attributes of their products. Successful evaluation under this program results in a Verification of Attributes Report™ (VAR™). The VAR™ can be helpful to qualify for points under major green rating systems or green codes and standards. The sustainable attributes are evaluated based on nine established guidelines. The current nine guidelines are:

- Evaluation Guideline for Determination of Recycled Content of Materials (EG101)
- Evaluation Guideline for Determination of Biobased Material Content (EG102)
- Evaluation Guideline for Determination of Solar Reflectance, Thermal Emittance and Solar Reflective Index of Roof Covering Materials (EG103)
- Evaluation Guideline for Determination of Regionally Extracted, Harvested or Manufactured Materials or Products (EG104)
- Evaluation Guideline for Determination of Volatile Organic Compound (VOC) Content and Emissions of Adhesives and Sealants (EG105)
- Evaluation Guideline for Determination of Volatile Organic Compound (VOC) Content and Emissions of Paints and Coatings (EG106)
- Evaluation Guideline for Determination of Volatile Organic Compound (VOC) Content and Emissions of Floor Covering Products (EG107)
- Evaluation Guideline for Determination of Formaldehyde Emissions of Composite Wood and Engineered Wood Products (EG108)
- Evaluation Guideline for Determination of Certified Wood and Certified Wood Content in Products (EG109)

The reports issued through the ICC-ES SAVE Program are similar to the concept of ICC-ES Evaluation Reports that have been used by the construction industry for several decades. These reports cover products that are alternatives to code-specified, provide independent evaluation of manufacturer claims, are broadly accepted by building officials, streamline the job of green verifiers, clarify environmental attributes of products and provide optional references to overall code compliance (i.e., structural, fire, durability, etc) via product ICC-ES Evaluation Reports.

The ICC-ES SAVE reports (VAR) are available online for access by anyone. A sample VAR is shown.

ICC EVALUATION SERVICE

Most Widely Accepted and Trusted

ICC-ES SAVE Verification of Attributes Report™

VAR-1053

Issued July 1, 2009
This report is subject to re-examination in one year.

www.icc-es.org/save | 1-800-423-6587 | (562) 699-0543   *A Subsidiary of the International Code Council®*

DIVISION 07—THERMAL AND MOISTURE PROTECTION
Section 07 21 16—Building Insulation
Section 07210—Building Insulation

REPORT HOLDER:

A–1 Insulation, Inc.
123A Rocky Road
Asphalt, CA 43210
(123) 765-4321
www.a1insulation.com
jv@a1insulation.com

EVALUATION SUBJECT:

A–1 Insulation

1.0 EVALUATION SCOPE

Compliance with the following evaluation guideline:

ICC-ES Evaluation Guideline for Determination of Biobased Material Content (EG102), dated October 2008.

2.0 USES

A–1 Insulation is a semirigid, low-density, cellular isocyanate foam plastic insulation that is spray-applied as a nonstructural insulating component of floor/ceiling and wall assemblies.

3.0 DESCRIPTION

A–1 Insulation is a two component system with a nominal density of 1.0 pcf (16 kg/m³). The insulation is produced by combining the two components on-site. Water is used as the blowing agent and reacts with the isocyanate, which releases a gas, causing the mixture to expand. The mixture is spray-applied to the surfaces intended to be insulated.

The insulation contains the minimum percentage of biobased content as noted in Table 1.

4.0 CONDITIONS

Evaluation of A–1 Insulation for compliance with the International Codes is outside the scope of this evaluation report. Evidence of compliance must be submitted by the permit applicant to the Authority Having Jurisdiction for approval.

5.0 IDENTIFICATION

The A–1 Insulation spray foam insulation described in this report is identified by a stamp bearing the manufacturer's name and address, the product name, and the VAR number (VAR-1053).

TABLE 1 – BIOBASED MATERIAL CONTENT SUMMARY

% MEAN BIOBASED CONTENT	METHOD OF DETERMINATION
15% (+/–3%)[1]	ASTM D6866

[1] Based on precision and bias cited in ASTM D 6866.
